American Football and the American Way of War

Daniel Sukman

American Football and the American Way of War

The Gridiron and the Battlefield

palgrave
macmillan

Daniel Sukman
Williamsburg, VA, USA

ISBN 978-3-031-55344-8 ISBN 978-3-031-55345-5 (eBook)
https://doi.org/10.1007/978-3-031-55345-5

This Palgrave Macmillan imprint is published by the registered company Springer Nature Switzerland AG
The registered company address is: Gewerbestrasse 11, 6330 Cham, Switzerland

Paper in this product is recyclable.

Preface

I've been lucky to be a member of several fraternities in my life. Not the Greek fraternities you find at a university or college, as my Alma Matta, Norwich University banned fraternities from campus sometime in the 1950s, long before my time. The fraternities I speak of aren't confined to college campuses nor are they composed of only men. In this sense, they become more of a tribe. I see the military as one of the fraternities. Often those I served with speak of the military, and even more specifically of the units I was a member of, as a brotherhood. Women are a part of this too, and brotherhood might be the wrong term to use, in a sense it is a tribe. Regardless of the term, through rigorous training and combat deployments, my brothers and sisters in arms and I forged eternal relationships that are impervious to our religious and political disagreements.

A second fraternity I was a part of was youth and high school football. I started playing football when I was 8, ending my time in competitive football two months shy of my 18th birthday. This is more of a fraternity in the strict sense as coaches and players were all boys and men. Looking back to my youth I see it as a fraternity, not in some Al Bundy never letting go of past glory, but as the method I used to create life-long friendships. I was lucky enough to grow up from kindergarten through high school in the same town, with the same friends who were the same teammates. We had a unified sports program and thus were more than classmates for over a decade, we were teammates.

A month past my 17th birthday, on a cold winter day in February of 1995, an army recruiter drove me to Fort Hamilton in New York City where I enlisted in the Army Reserve. It was February 14th, Valentine's Day, which served as an omen of what lay ahead. I would leave for basic training that summer between my junior and senior years of high school. While my friends and teammates were taking driver's education, I was living in the barracks at Fort Jackson, waking up at 4:30 in the morning to run and do push-ups, followed by afternoons on the rifle and bayonet range. From the first morning wake up that summer I could see the similarities between football and the military and have felt ever since that playing football in my youth prepared me for adulthood in the army.

Fig. 1 Football Players lined up in a military like formation. (*Source* Image in the public domain from Wikimedia Commons)

Morning physical training (PT) began with a rectangular formation of soldiers spread out two arm distances apart stretching, doing push-ups and sit-ups and some other calisthenics much in the same way football practice would start, only instead of the team captains leading the group it was the drill sergeants. This morning ritual occurs on every army base every morning with non-commissioned officers (NCOs) leading their soldiers, ensuring they will be fit enough when the time comes for combat.

Then there is the uniform. Within the first week at Fort Jackson, I was issued my camouflage battle dress uniform (BDU). There is something about wearing the uniform that makes you more conscious of a feeling of belonging. There is a sense of pride that comes with shined boots and a starched uniform in garrison, and all the same for dirty, muddy, wrinkled, and torn uniforms while in the field training for or during combat. Indeed, wearing a uniform is something soldiers have done for generations. It's a way to build cohesiveness, but more practically a way to tell the good guys from the bad guys on the fields of combat. British soldiers didn't wear red coats to mark themselves as easy targets, they wore bright red so that they wouldn't get shot by their side on the smoke-filled battlefields in the days of the musket. Army uniforms have a nametag on the front, but football uniforms have the name spelled out on the back. Everyone gets their own number which is one of the few ways of individuality on the team. But the football uniform, like a military uniform, gives one a sense of belonging. Every Friday during the fall football season was a day when members of the team would wear the jersey to school. This ritual began when we were eight years old and continued through our senior year of high school having no other purpose than to show unity. Unity among teammates and friends (Fig. 1).

From the time we played Peanuts (8- and 9-year-old) in Pop Warner through high school, we had various cohorts of coaches that we all looked up to for guidance and wisdom. A football coach's word is the unquestioned law on the gridiron. Kids sometimes have an oversized view of their own experience, and I was no different, often thinking that the accumulated experience of teachers and other adults wasn't worth very much. This belief stopped on the sidelines and for the most part, players hung on to every word that our coaches spoke. Not only did we think they had the key to victory for every game, but they were always looking out for our best interests. Loyalty was a two-way street between coaches and players. Those I played for were stand-up men. I saw the same interpersonal dynamics in the army, and for the most part still see this today. Junior soldiers look up to their leaders and follow orders with the understanding that the NCOs and officers leading them have their best interests of them, and the best interests of the unit at the forefront of their minds in every decision.

As my football days peaked in high school, so did my social activities. House parties when parents were away on vacation were a common event where the football team would get together and drink cheap beer. If no house was free, we would have bleacher parties where we would sit on the bleachers at the football field, chug beer, and talk about the things that teenagers talk about. And when the police chased us away from the bleachers, we would sit on the Bethpage Golf Course, the same one where Tiger Woods would win the U.S. Open a few years later, to carry on our buzz-fed conversations. There were no sprints, tackling drills, or scrimmages, but the social bonding that occurred was just as important to team chemistry as our end-of-practice wind sprints.

To earn my college degree and my commission as an officer in the U.S. Army, I spent four long years in the Corps of Cadets at Norwich University, a small school in the heart of Vermont's Green Mountains. I didn't play football, and spent my time there focused on the classroom, R.O.T.C., and Corps of Cadet activities. I embraced the atmosphere, the comradery, the leadership opportunities, the self-discipline, and the hard work it took to succeed at a military school. Morning formations meant hangovers weren't an option, and by my sophomore year, I understood that my performance in the classroom would have an impact on the rest of my life. I buckled down, studied hard, improved my grade point average, and was awarded a two-year R.O.T.C. scholarship which paid for school and came with an eight-year commitment to the army. Much like my time in basic training where I saw athletes embrace the teamwork it took to succeed, I saw the same dynamic at Norwich. It is one I have observed in every assignment since.

When I was commissioned as a second lieutenant upon graduating from Norwich University, I would attend the officer basic course at Fort Leonard Wood. There was a social event at the officer's club nearly every week where senior officers on the base would gather, drink beer, and tell war stories about the various assignments and deployments over their careers. The lieutenants would gather around our tables and talk or listen silently as small crowds gathered around the colonels hoping for a glimpse into their collective wisdom. The environment at the officer's club often lubricated by alcohol lowered the formality of rank allowing for a true dialogue of our thoughts and ideas. Just like our bleacher parties, I understood the importance of how social gatherings can improve the collective performance of a team whether the team wore a football or military uniform.

Listening to the senior members of the football tribe is where I first learned how a chain of command works. While the trend in business is to grow flat organizations, the military is still hierarchical. Within units, rank still matters and signifies the level of responsibility and or command someone holds. Military units require subordinates to follow the lawful orders of their superiors be they NCOs or officers. Today's culture encourages more dialogue between ranks and organizations, and that is paramount for bringing out the best ideas, and in many cases preventing the worst ideas from coming to fruition. However, just as there is no room for debate in the huddle, in combat there isn't always the time to debate a course of action nor to vote on which direction to move. Trust is earned, but there is always some amount of faith that the senior leader will do what's best for the soldiers and for the squad, platoon, or company.

On the football field, especially in youth and high school ball, the word of the coaches was law. Certainly, they were open to ideas as they came about, but during a game, there was no time to debate a play call with the coaching staff. You do what you're told, execute the game plan, or be prepared to be replaced by the kid sitting on the bench itching to take your place. As a kid in youth and high school ball, we had the utmost confidence that our coaching staff knew what they were doing (and looking back for the most part they did). As I have moved from job to job and base to base across the globe, I see that soldiers who can instinctively follow the chain of command are often alumni of football or other team sports. This is of course a generalization based on my anecdotal experience and soldiers who played individual or no sports at all can follow the chain of command like anyone else.

Football taught me a lot about failure which is useful having served in a military that fought largely unsuccessful operational and strategic campaigns in Iraq and Afghanistan. I go back to my last years in Pop Warner football when our Farmingdale Hawks team and my fellow cohort of athletes won one game over two years. Most games were complete blowouts where we lost by 30, 40, and on occasion 50 points. Going 1–15 is humbling but generally prepares you for a life where we often fail more than we succeed. I failed a lot in this life, in athletics, in work, and in love but like those two years of football I didn't let failure stop me from trying again. After those horrendous losses, we showed up to practice the next week ready to put in the work and try our best in the next game. Other teams had more talent, but I look back and know that they probably didn't learn as much as we learned about ourselves and what we learned about each other.

Throughout my deployment to South Baghdad from 2005 to 2006, I experienced failure and defeat as a member of the U.S. military. I was part of the second brigade combat team, (502nd), in the 101st Airborne Division. Jim Frederick wrote about our time there in the book *Blackhearts*, which was not kind to our collective combat performance. Within the brigade, a few soldiers committed atrocious war crimes which is, to say the least, not helpful when trying to conduct counterinsurgency operations. Throughout the year-long deployment, we lost 55 soldiers killed and scores who would earn a Purple Heart. The casualty count continued when we came home, as I lost a mentor to suicide. The effects of PTSD remained long after we arrived home. Above us, Division and Corps and above echelons were just as lost making what seems now one poor decision after another. When I redeployed to Fort Campbell, I contemplated leaving the army for the civilian world. Turns out I didn't, I drew on what I think was some of the strength I learned from those two years of constant defeat and stuck with the uniform. I'm glad I did, and so far, it seems like it was the right choice.

Nine years into my career and shortly before pinning on major I was accepted into the Army Strategist career field. The application of what I learned from football expanded into the windowless planning rooms where I helped to conceptualize, develop, and write war plans. No individual is responsible for the entirety of a war plan, which includes annexes and appendixes that can exceed well over 500 pages. Rather, as a planner I had to work with groups of individuals, each with their specific knowledge, skills, and abilities. I had operational planners, planners for logistics, information operations, communications, and intelligence all contributing their part. Planners came from each service and often included people wearing uniforms and suits. Essential to writing these plans was ensuring everyone's thoughts and ideas complimented each other and worked toward the same objectives and policy goals. Should one person fail in their planning, the strategy could fail spectacularly. Writing strategies and operational war plans is indeed a team sport.

Not only did I have the privilege to write war plans, but I also had the opportunity to teach strategy and planning for three years at the Joint Forces Staff College in Norfolk, Virginia. Here I stood in front of seminars composed of senior officers who were looking to understand the nuances of joint doctrine, how to develop and compose strategy, and how to lead planning teams at the operational echelon of war. Again, similarities to football emerged, only this time from the perspective of coaching. Teaching and coaching are similar, in that both require the leader to train and build the expertise, knowledge, and skills of those they lead.

I left the Staff College in 2022 for my first assignment in the Pentagon as the Chief of Strategy on the Joint Staff. This was my capstone assignment and one that offered me deep insights into how the Joint Force and the Defense Department operated. I was responsible for national level strategic documents such as the *National Military Strategy*, the *Unified Command Plan*, the *Chairman's Risk Assessment*, and a host of other documents that would gain signature by the President, the Secretary of Defense, and the Chairman of the Joint Chiefs of Staff. I witnessed the interactions and relationships between senior uniformed and civilian leaders and had to use my position, experience, skill, and dare I say wisdom gained from over two decades in service to craft strategic guidance that had the concurrence of well over a dozen 4-Star officers, the Secretary of Defense, and the President. Sitting in the room with the senior military leadership of the United States, I often reached back to the skills and abilities I started learning as an eight-year-old kid.

Football was where I first learned the benefit of a team effort. While leading those planning teams I was able to reach back to my football days of youth where players of various skills playing at different positions were asked to execute their part of the game plan, week in and week out. Should one offensive lineman miss an assigned block, or a receiver run the wrong route, a play, a series, and a game could be lost or result in an injured teammate. History is replete with quarterbacks going down because of a missed block or misunderstood assignment. Everyone does their part, and if you don't do yours with enough consistency somebody else is waiting to do it for you. Nobody is indispensable. The tribe acts swiftly to find a replacement.

More than the team effort I came to see parallels between my chosen profession and the concepts that go into football. Football isn't war, and some people get upset at the comparison, but there is no other sport that comes close. At every level of the game, there are similarities. Indeed, when I try to explain planning and operational concepts to non-military professionals, I reach back to what I learned playing football to describe them. This book is the result of years of thought and countless hours translating those thoughts onto paper.

The Chapters

Chapter 1 is an introduction and details some of the previous scholarship and writings on the confluence of football and war. The chapter traces some of history of the war and football analogy, the very heart of this book.

Chapter 2 will cover some of the history of football and how it relates to the military. It includes backgrounds of famous presidents and generals who played the game and details famous coaches and players who served in uniform. Moreover, the introduction offers the perspective on just how large of a monolith football is in American society. The history of the U.S. military and American Football are interrelated, and in my research, I was surprised to discover that many of America's national heroes spent their formative years playing, coaching, and studying the game.

Chapter 3 is an overview of international relations theory that centers on realism and liberalism. It defines the various theories such as realism and liberalism then explains how they apply in real-life geopolitics and how they apply to football. Nations and teams cooperate, and nations and teams compete, all at the same time. There are stark similarities that help to explain some complicated subject matter and reveal the broader aspects of the game that occur outside the lines.

Chapter 4 discusses the foundations of strategic theory through the lens of history's great strategic thinkers from Clausewitz, Thucydides, and Sun Tzu to more contemporary theorists like Alfred Thayer Mahan and his thoughts on war in the maritime domain, J.C. Wylie and his idea of control, and John Boyd's OODA Loop. Theorists don't necessarily provide a solution to war but rather provide a method to think about war. These theorists describe what political and military leaders should consider when choosing a strategy to employ. Further, it is theorists who describe the conditions of war from the tactical through the operational and to the strategic level. Many of these conditions apply to the game of football.

Chapter 5 is where we apply the various thoughts of strategic theory to specific applications. From attrition and exhaustion to the complicated matters of deterrence, nations, militaries, and football teams alike must decide on the overall theory that will lead to victory. This chapter helps to explain each strategy with some historical examples of the application of these strategies.

Chapter 6 covers U.S. military doctrine and bins war and decision making into three distinct levels of war; strategic, operational, and tactical. Chapter 5 will introduce and define these levels of war, provide examples of activities in each level, and relate each level of war to its football equivalent. War is more than what is happening in a specific geographic space, and football is more than what happens between the sidelines. It is important to understand the various echelons or levels and where each part of a broad organization fits into this paradigm.

Chapter 7 describes the principles of war. At the operational level of war, the U.S. military uses specific principles of war and the principles of joint operations to guide its warfighting doctrine. These principles include objective, offensive, mass, economy of force, and surprise to name a few. Chapter 7 will explain each principle and how they apply to war and the game of football. While not a checklist for success, thinking about and applying principles to strategies and plans is a useful construct in the development of a theory of victory. Moreover, there are times when a leader must assume risk in one or more principles to exploit and take advantage of others.

Chapter 8 is all about doctrine. Doctrine is the foundation upon which militaries function. Chapter 8 will discuss the importance of U.S. Joint Doctrine, how doctrine bins various capabilities, and relate the same practice to the application of doctrine in the form of playbooks and other methods in football. Chapter eight uses the lens of the seven joint functions of movement and maneuver, intelligence, fires, sustainment, protection, command and control (C2), and information.

Chapter 9 will examine design and problem solving. At the operational and strategic level of war, commanders and staffs employ a design methodology to solve complex and complicated problems. This methodology contains specific elements that lead military planners to the formation of an operational plan. At the tactical level of war, planning is often simple with directed guidance from higher headquarters, but at higher echelons, staffs and commanders must recommend to their military and civilian leaders what it is they should seek to accomplish and create the methods in which they will attain success. How military organizations do this is part systematic and part creativity.

Chapter 10 covers innovation and evolution in war and football and the conditions that force major changes in the way wars are fought. From new and emerging technology to the concepts of visionary leaders. The chapter discusses these ideas through the evolution of football from its early history to the current day. This chapter concludes with a discussion on what prevents organizations from innovating. Organizations continually look for ways to innovate, evolve, and adapt to their circumstances. Football teams and military organizations alike must stay one step ahead of their adversaries and competitors and any small advantage can turn out to be decisive in victory.

Chapter 11 centers on civil-military relations. In America, it is gospel that those in uniform are subordinate to elected civilian leaders. The chain of command runs from the President to the Secretary of Defense, through senior military commanders. For America's National Guard, the chain of command runs up through each state and territorial governor. Similar relationships occur across the spectrum at each level of football including owners, coaches, players, university presidents, school boards, and parents of youth players. The chapter then delves into various cultural conflicts within the United States, and how the military and football's policies tend to mirror broader society.

Chapter 12 discusses the similarities that apply at the individual and organizational level between war and football, and Chapter 13 is where I discuss the differences. Despite all the similarities between war, the military, and football, there are stark differences. From the use of alliances and coalitions, to agreed upon rules, football and war are not a 1-1 translation. Chapter 13 will spell out these differences for the reader as a reminder of the harsh realities when nations send their young men and women into combat.

This book will be the first writing to describe the intersection of football and the American way of war. This book is the first writing to detail American doctrine and warfighting concepts, and to use football as an analogy and broad descriptor of this method of war.

This book will take the reader on a wild ride with descriptions of military strategy and tactics, the Xs and Os of football, and use a bit of military and football history to explore the American way of war, which reflects American society. War and football each have the potential to bring about the very best and very worst in people. War and football lead people to the highs of victory and the agonies of defeat.

Williamsburg, USA Daniel Sukman

Acknowledgments

"All that we are began in our youth"

This book is the culmination of eight years of thinking and writing on the topics of football, war, strategy, and planning.

To my Farmingdale Hawks coaches and teammates, I still owe a debt of gratitude to the coaches who shaped me in my youth to become the person I am.

To all my colleagues I encountered during my time in uniform over the past 29 years, thank you. Soldiering is not what I do, it is what I am.

For all who supported my writing, through encouragement, editing, and publishing, thank you for giving me a voice.

My family has always supported all that I do. Thank you, Mom, Dad, and Matt for your unconditional love.

Thank you to Jill for always supporting my endeavors and to Sabrina and Brendan who give me purpose in this life. I love you all so very much.

Book Information Text

This book provides the reader with a history on the linkages between American Football and the American Way of War. It provides concise summaries of aspects of tactics, operations, and strategy, as well as international relations theory using football as a metaphor to simplify the concepts. International relations theory and military strategy tend to be complicated subjects often explained with jargon that puts the average reader to sleep. A book about war and foreign policy explained through football analogies and comparisons is designed to liven the topic and explain various military theories, strategies, and concepts. War and football tend to dominate national media. War is often a topic of debate from the highest levels of government to local bars and coffeehouses. Football is the monolith of sports that blankets sports television and radio throughout the year. This book provides the reader an understanding of how military professionals formulate and execute strategy in an easily understandable manner.

Contents

About the Author

Daniel Sukman was born and raised in Farmingdale, New York. He is a graduate of Norwich University and holds graduate degrees from Webster University, Liberty University, and the U.S. Army War College.

As a U.S. Army officer, he now serves as an Army Strategist. He currently works in the Pentagon and on the Joint Staff as the Chief of Strategy Development, Joint Strategic Plans, J5. Previously, he taught strategy and planning at the Joint Forces Staff College, served as a planner putting together numerous war plans and strategies for Combatant Commands as a member of the Joint Enabling Capabilities Command. He also served as a crisis action planner at European Command. He further wrote and contributed to Joint and Army Concepts while serving at the Army Capabilities Integration Center. He has served in planning assignments in both the institutional and operational force and has combat deployments to Iraq and Afghanistan. At the tactical level, he served as a company commander in 2nd Brigade of the 101st Airborne Division.

He is married Jill in November of 2008, they have two children, Brendan and Sabrina.

He is an avid sports fan, consistently cheering for the New York Giants, Mets, Islanders, and Knicks.

How it started: The author with his youth football team. Dan is #60, third row next to his coach

How it's Going: The author with members of the Joint Staff and Chairman Mark Milley. Dan is just off center, second row from the top

Abbreviations

AFL	American Football League
BRAC	Base Realignment and Closure
C2	Command and Control
CCIR	Commander's Critical Information Requirements
COG	Center of Gravity
COIN	Counter Insurgency
DADT	Don't Ask Don't Tell
EBO	Effects Based Operations
FFIR	Friendly Forces Information Requirements
FOB	Forward Operating Base
HASC	House Armed Services Committee
HBCU	Historically Black Colleges and Universities
HUMINT	Human Intelligence
IAAUS	Intercollegiate Athletic Association of the United States
IDF	Israeli Defense Forces
ISIS	Islamic State in Syria
ISR	Intelligence, Reconnaissance, and Surveillance
MWR	Morale, Welfare, and Recreation
NATO	North Atlantic Treaty Organization
NCAA	National Collegiate Athletic Association
NFL	National Football League
NWFL	National Women's Football League
OODA	Observe Orient Decide Act
PIR	Priority Intelligence Requirements
PTSD	Post-Traumatic Stress Disorder
R.O.T.C.	Reserve Officers' Training Corps
RPO	Run Pass Option
SAMVA	Special Assistant for the Modern Volunteer Army
SEC	Southeastern Conference
TBI	Traumatic Brain Injury
UCLA	University California Los Angeles

UCP	Unified Command Plan
USC	University of Southern California

List of Figures

Chapter 1
Introduction

This is not the first writing to compare football and war. Since the founding of the game in the late 19th Century, there has been an abundance of writings that compare what happens on the football field to what happens on the battlefield (Fig. 1.1).

From the game's beginnings, there are countless articles, essays, and dialogues on how football relates to war. These writings include those of Walter Camp, known as the founding father of Football. Indeed, Camp compared football and war to *The New York Times* in 1917. In his description, Camp stated "The gridiron is the battlefield, the two teams are the opposing armies…"[1] Camp further describes the intricate planning, the discipline of team movements, and the necessity for creativity and imagination that enables success in football and war. This book pays homage to Walter Camp with the title; *The Gridiron and the Battlefield*.

In *Football for Public and Player*, Herbert Reed wrote one of the earliest complete comparisons of football and war.[2] Published in 1943 the book provides a detailed description of the rules and strategies of the game. Chapter two, however, examines the similarities and differences between war and football as both were understood and practiced at the time. The book tends to show its age by limiting the discussion of war to the strategic and tactical levels, linking them directly without the modern understanding of the operational level. The author does conceptualize the difference between battles and campaigns, the latter of which is a series of interconnected or interrelated battles. From the football perspective, the author highlights the importance of the kicking game, with special emphasis on punting. In the modern era, the kicking and punting game while important is not on the same level of importance as the passing and running game of the offense.

In 1939 W. Glenn Killinger published his book *Football*. Chapter 9 of *Football* titled "Generalship and Strategy" refers to the quarterback as the field general of the team, one who is responsible for the team's strategy, and one who all other players look to for leadership on the field. Written in the day when quarterbacks called their own plays, Killinger points to several traits a successful quarterback must possess to succeed, including playing the opponent's weakness and thinking under fire.[3]

D. Sukman, *American Football and the American Way of War*,
https://doi.org/10.1007/978-3-031-55345-5_1

Fig. 1.1 Walter Camp, considered the founding father of professional football (Image from the public domain from Wikimedia Commons)

While the book does not make a one-to-one comparison to war, the leadership traits Killinger describes will be familiar to American military officers.

In 1965, *Esquire Magazine* published Thomas B. Morgan's essay "The American War Game," an article that details the violence of the sport and compares the protective equipment of football players to the body armor worn by soldiers throughout history on the battlefield.[4] Morgan understood that the defining characteristic of war is violence, an insight gleaned by historic military theorists such as Carl von Clausewitz.

Violence and brutality, especially before modern safety initiatives took hold was a central feature of football in the mid-twentieth century.

In the book *Sports in Modern America,* Douglas Noverr and Lawrence Ziewacz acknowledge Morgan's insights and explain the conundrum that Americans tend to be inconsistent with their views towards violence. On the one hand, Americans condemn uncontrolled violence but tolerate and even extol violence that occurs within the bounds of a game.[5] Indeed, televised football and highlight videos often showcase the most violent of tackles. The authors also contend that unacceptable violence is quickly lost with the focus on the following week's game. From warfare to football, violence is permissible within the written and unwritten rules of the contest.

Multiple authors described how various wars impacted the game. In his book *War Football,* Chris Serb detailed the history of service football teams in the First World War. Indeed, this history accounts for the myriad of teams that formed to entertain the troops and to provide competitive recreation activities at newly populated military installations housing the millions of service members conscripted into the force. Serb's book properly recognizes how World War I and service teams combined to introduce millions of people to the sport, and how service teams made it possible for former college stars to play after graduation, as professional football had yet to take off in the way professional baseball already had.

In a more specific or concentrated work *Fields of Friendly Strife,* sports historian Timothy Brown detailed the 1918 and 1919 Rose Bowls, games played by service teams as colleges could not field or send teams to compete during the height of American involvement in World War I. Looking past the First World War, Russ Crawford expanded the history of the military and football and brought it through the modern era in his book *Le Football.* This book centers on the history of football in France, which relies heavily on service teams beginning in World War I. Le Football accounts for the myriad of "Bowl Games" played on the battlefields of France and Algeria (a French Colony at the time) and how the war and the military brought the sport to millions who otherwise would never have seen a game.

The military and football have come closer in their relationship in recent years. In *Super Bowl Monday,* Adam Lazarus describes the physical and emotional environment that surrounded Super Bowl XXV, a game played with the 1991 Gulf War (Desert Shield / Desert Storm) occurring in the background. Lazarus tells the story of how for a moment, there was an argument on whether to cancel the Super Bowl with the nation at war. The game went on as scheduled, albeit with an intense security atmosphere and overwhelming displays of patriotism. The book is a reminder of just how big of a role football plays in modern American culture.

The comparisons of the language and jargon are not new either. In 1988 James Weeks wrote the short essay "Football As a Metaphor For War" for *American Heritage.*[6] In the essay, Weeks reaches into the late 19th Century to examine how the early players of the game immediately sensed how military language and metaphors could best describe the game. Weeks's research reveals how following the Civil War football became a method for young men to compete and develop their character in a way that would be absent when these young men were not on the battlefield.

In 2009 Michael Oriard, the author of *Brand NFL* wrote a piece for Slate describing the nexus between the National Football League and war. Oriard details how the jargon of war and football crosses from the generalship of quarterbacks to military operations such as Operation Varsity in WWII and Operation Linebacker in Vietnam. Further, Oriard describes how football fields became a place of hyper-patriotism that includes flyovers, salutes to the flag, and patriotic themed pre-game shows.[7] Oriard in his writing understands that war is more than just the conflict between two armies on the battlefield, but rather an event that encompasses the whole of a nation and its society. Militaries don't win or lose wars, nations do.

Deep into America's Global War on Terror, prolific writer and author Mark Edmundson published an editorial in the Los Angeles Times in 2014 titled "Football is America's War Game." In this writing, Edmundson declares that the warlike culture of America was bound to embrace a warlike sport.[8] Reflectively, Edmundson rattles off the wars America fought throughout the 20th and into the 21st Century taking the reader to the logical conclusion that America is a violent nation, more often at war than at peace in the world. It is then logical for a warlike nation to adopt football as its favorite sport, perhaps even replacing baseball as the regarded national pastime.

It's not only books and magazine articles that compare football and war. Famous military and political leaders have a long history of comparing the two, often using them as analogies to describe either one. Douglas MacArthur famously said, "Upon the fields of friendly strife are sown the seeds that on other days, on other fields, will bear the fruits of victory," when discussing the importance of football and other sports at the academy. The inscription on a plaque at West Point's football stadium reads, "If I want an officer for a secret and dangerous mission. I want a West Point Football Player." This quote is attributed to General George C. Marshall.

The links between football go well beyond a couple of quotes from our 5-Star Generals of World War Two. The next chapter will expand the knowledge of the deep interdependence of America's two most popular topics.

Notes

1.. "Likens Football to War. Walter Camp Compares the Gridiron to the Battlefield." *New York Times*. 4 November 1917, pg 37.
2.. Herbert Reed, *Football for Public and Player*. (New York, NY: Frederick A Stokes Company Publishers, 1943).
3.. W. Glenn Killinger, *Football*. A.S. (New York, NY: A.S. Barnes and Company, 1939), 137.
4.. Thomas B. Morgan, "The American War Game." *Esquire Magazine*. October 1965. 71–72, 142, 146–148.
5.. Douglas A. Noverr and Lawrence E. Ziewacz, "Violence in American Sports," in *Sports in Modern America*. ed. William J. Baker and John M. Carroll. (St. Louis, MO: River City Publishers, 1983), 138–139.
6.. James Weeks, "Football As A Metaphor For War." *American Heritage*. September/October 1988. Retrieved from: https://www.americanheritage.com/football-metaphor-war

7.. Michael Oriard, "Flag Football: How the NFL became the American War Game." *Slate*. 17 November, 2009. Retrieved from: https://slate.com/culture/2009/11/how-the-nfl-became-the-american-war-game.html
8.. Mark Edmundson, "Football is America's War Game." *Los Angeles Times*. 23 August, 2014. Retrieved from: https://www.latimes.com/opinion/op-ed/la-oe-edmundson-americas-and-war-connection-20140824-story.html

Chapter 2
The American Military and Football

> The Supreme quality for leadership is unquestionably integrity. Without it, no real success is possible, no matter whether it is on a section gang, a football field, in an army, or in an office. – Dwight D. Eisenhower.
>
> In life, as in football, the principle to follow is to hit the line hard. -Theodore Roosevelt.

Baseball is America's pastime. Football is America's sport. Football is a part of life in nearly every community in the United States. From densely populated cities to the open plains of rural America, football finds itself a central part of life. The sport owns national television ratings with contests from high school through college to the National Football League (NFL). Showcasing Football's cultural dominance in high school and college are movies like *All the Right Moves, Remember the Titans, Necessary Roughness, The Program, and We Are Marshall* and the popular television series *Coach*, and *Friday Night Lights*. On Autumn Saturdays, sold-out college stadiums populate the American landscape.

Nowhere else in the world can one see college students cheering, roaring, and doing everything else—some unspeakable things—like at American college football games. On Sundays, NFL games fill the day, even spilling into Mondays and Thursdays. During the height of the COVID-19 pandemic, games continued in empty stadiums and appeared on Tuesdays and Wednesdays, capturing the highest television rating each night of the week. Super Bowl Sunday has become the closest thing to a national holiday celebrated across demographic groups as a day can get. As former NFL commissioner Paul Tagliabue once stated, "the Super Bowl is the winter version of the Fourth of July."[1]

As for official federal holidays in America, it has become a Thanksgiving tradition to watch the Lions and the Cowboys play in the evening following the family meal. New Year's Day is dedicated to college football with the major bowl games scheduled for January 1st. Accompanying New Year's Day bowl games are large parades with floats where tens of thousands of spectators line the streets to celebrate the game. For many communities, games are more than just the four quarters on the field, rather a weeklong event with bonfires, pep rallies, parades, and events celebrating school spirit in the prelude to the game. Football games capture everyone's imagination and let loose their "Monday morning quarterback." Hardly anything in the week goes

D. Sukman, *American Football and the American Way of War*,
https://doi.org/10.1007/978-3-031-55345-5_2

without commentaries on the airwaves and at the table without being related to this glorious cultural phenomenon.

Professional football games generally occur on Sundays, the Sabbath for those who practice Christianity. While mega-churches can attract thousands of worshippers on a Sunday morning, NFL stadiums pack tens of thousands into their seats and represent a true cross-section of America with multiple races, religions, and ethnicities coming together to cheer for a common cause. The community extends beyond the stadium to packed sports bars where strangers raise their glasses and hoot and holler when their team scores a touchdown or sacks the opposing quarterbacks. Football creates virtual communities where friends compete against each other online playing fantasy football. These community gatherings happen in college towns on Saturdays and in small towns and suburbs across the nation when family and friends watch their teenage sons play football under the Friday night lights. Football is polytheism and its idols are plastered on walls, televisions, and computer and smartphone screens across America.

Football dominates the television market like no other sport or show. Regular season Sunday, Monday, and Thursday night games consistently place #1 in the primetime Nielsen ratings vastly surpassing every show on every network. Even an event as mundane as the NFL draft draws large viewing audiences. Indeed, in 2017 more than 200,000 people stood in downtown Philadelphia to watch the annual draft.[2] While in some nations gatherings this large represent large scale protests against a sitting government, only in America will 200,000 people brave the elements to watch people announce the names of soon to be professional football players. In 2015, the television audience for the NFL draft topped 45 million people.[3]

Football is not just a sport; it is a part of American culture, and more importantly, a vast part of American strategic culture that deals with the American way of national security policy, the justification for the use of force, and the conduct of war. Analogies are useful in helping to provide simpler explanations to complex ideas. Football serves as a useful analogy—a tool—to explain the American way of war.

The relationship between football, the American Military, and war begins with the Navy song *Anchors Aweigh, which* began as the Naval Academy's football march. But more than fight sons, football analogies, comparisons, and jargon easily make their way into American military culture.[4] Examples include Operation Goalpost,[5] and Operation Varsity[6] and "Operations Linebacker and Linebacker II in the Vietnam War.[7] Other terms and phrases such as Red Zone, and inside the trenches are used to describe situations on the football field as on the battlefield. Every play on the gridiron begins with each side lined up in a set formation conjuring up a mental picture of historical battles where armies lined up across an open field in the traditional way of war from Marathon to Gettysburg, and many of the battles in between. Each year in the offseason, scouts, head coaches, and general managers of NFL teams select players from the college ranks in the annual draft from what they designate as their "War Room," a self-comparison to the nation's Commander In Chief.[8]

American football came with the rise of the United States as a great power and eventually a world leader. The evolution and transformation of both the sport and the nation have gone hand in hand. While a powerful nation breeds a masculine sport like football, this game has come back to influence the way Americans, especially American leaders, think and behave. To an extraordinary extent, the United States, and the U.S. military, behave like a football team forcing its way through international politics and making massive impacts everywhere. Nations tend to view their place in the world with the United States as a comparison.

As to the title of Commander In Chief, since 1900, eight American Presidents played football at some level in their lifetime. General of the Army and later President Eisenhower played running back and linebacker while studying at West Point. John F. Kennedy played at Harvard, while President Richard Nixon was a backup tackle at Whittier College. In 1934, President Ford, perhaps the best athlete to serve as the Commander in Chief was the team MVP for the University of Michigan.[9] Ronald Reagan played at Eureka College long before he earned the nickname "The Gipper" in playing George Gipp in the 1940 film *Knute Rockne, All* American.[10] President Trump played football in high school at the New York Military Academy. President Donald Trump's influence in football began long before his political career as the owner of the New Jersey Generals of the United States Football League.[11] Joe Biden played football in junior high and high school as an end and halfback and credits the game for giving him confidence that he could do things he once believed he could never do.[12] For each of the men, regardless of the highest level they played, football was a formative event of their youth.

Outside of playing football, other former commander in chiefs played an outsized role in the game. Football was a key factor in Teddy Roosevelt's rise as a politician. When he served as New York City Police Commissioner, he was influential in bringing back the annual Harvard versus Yale game following its cancellation.[13] Teddy Roosevelt was a passionate football fan, having attended the second Harvard-Yale football game in 1876 where he watched Harvard lose 1–0.[14] While in office, President Teddy Roosevelt was witness to the sheer violence of the early game and mandated that colleges across the country figure out a safer way to play or otherwise face an end to the game. Following the 1905 college football season, President Roosevelt brokered a conference among collegiate coaches at the White House to institute new rules including legalizing the forward pass, increasing the distance for a first down from 5 to 10 yards, and establishing a neutral zone between the offense and defense, which became known as the line of scrimmage. Following the rule changes there was an instant decline in serious injuries and fatalities on the field. President Roosevelt's son, and future army general, Theodore Jr. played football for Harvard, and it was his injuries that led to his father leading the push for the rule changes.[15]

Before Teddy Roosevelt became President, he served as second in command of the famous Rough Riders in the Spanish American War. While filling out the regiment, Roosevelt and his commander Leonard Wood recruited former college football players and other athletes.[16] Roosevelt and Leonard Wood saw eye to eye on the value of football players as Leonard Wood once played and coached football for Georgia Tech. Leonard Wood is one of four graduates of Georgia Tech to have been awarded the Medal of Honor.[17]

Other Presidents who were involved in football before taking the oath of office include President Woodrow Wilson. Though he never played, he was an assistant coach to the Wesleyan football team. While a student at Princeton, Wilson wrote sports editorials for the campus newspaper often focusing his writing on athletics and football.[18] When there were debates and national discussions on banning the game due to its violence, Wilson became a staunch supporter of football, often defending the game in public debates. In a smaller role, Herbert Hoover worked as Stanford University's football team as a business manager and even worked at the gates collecting ticket receipts at Stanford's first ever football game in 1892.[19] Hoover's football business acumen, however, would not help cure the recession during his administration.

For General of the Army and later 34th President of the United States Dwight D. Eisenhower, football was central to his military career. According to his biographer Carlos D'Este, Eisenhower viewed an appointment to West Point to continue his pursuit of athletics, specifically football and baseball. Eisenhower played linebacker for Army and at one point lined up across Jim Thorpe when Army famously played and lost to The Carlisle Indian School in 1912.[20] Legend holds that a knee injury sustained in this famous contest would end Eisenhower's playing career but would open doors to the intellectual side of the game.

Eisenhower remained on the team and continued to coach teams throughout his time as a junior officer following his graduation from the Academy in 1915. Indeed, Eisenhower's early assignments often came at the request of senior officers who wished for Eisenhower to coach the post team.[21] For example, a mere four years removed from West Point, a young Lieutenant Eisenhower coached St. Mary's College, a school just outside of Fort Sam Houston to a winning record.[22] Eisenhower would later coach the III Corps football team during the interwar years which featured a game in 1922 against their heated rival Quantico in front of 60,000 fans, which included over one hundred members of Congress, the Secretary of War, and the Secretary of the Navy.[23] Without football, the United States doesn't get Dwight Eisenhower.

One of Eisenhower's teammates from the famous 1912 West Point team was then cadet Omar Bradley. Bradley was a backup lineman and a member of the 1914 undefeated and national championship West Point team along with his fellow teammate and future 4-star general James Van Fleet. All of them witnessed the famous 1913 upset loss to Notre Dame who used the event to transform the employment of the forward pass. Bradley would later play for the Camp Dodge service team during World War I and displayed much improvement from his cadet years by earning playing time as a starting tackle (Fig. 2.1).[24]

Eisenhower and Bradley's fellow 5-Star General of the Army Douglas MacArthur was a passionate football fan and played for his high school, West Texas Military Academy. Although too small to play for West Point as a cadet, in 1903 MacArthur still managed to serve as the team's student equipment manager, and week in and week out stood on the sidelines with the players and coaches.[25] Following World War I, MacArthur, who understood the relationship between sports and martial life returned to the Academy as Superintendent and implemented the intramural sports program that persists to this day.

Fig. 2.1 1912 West Point Football Team which included Dwight Eisenhower (third from left) and Omar Bradley (second from right). Photo from Wikimedia Commons

George S. Patton was a backup for West Point's football team throughout his four years at the academy. His playing was intense as he broke both of his arms in practice, but it was never enough to earn a starting position. Histrionics may be useful in earning battlefield fame, but not helpful in earning a starting position on a football team. Indeed, Patton was a better horseback rider than a football player, competing and taking fifth place in the pentathlon at the 1912 Stockholm Olympics.

Eisenhower Bradley, and Patton were not the only senior military officers with football backgrounds. James Van Fleet, who commanded a battalion in the First World War and a regiment, division, and corps in the Second World War, followed by commanding Eighth Army and United Nations forces in Korea—replacing Matthew Ridgeway, coached multiple teams in the interwar years to include Kansas State and the University of Florida.[26] Indeed, while working as a R.O.T.C. instructor at the University of Florida he also served as head coach of the Gators football team. The former fullback at West Point was an exceptional coach leading the team to a 12-3-4 record over two seasons.[27] Van Fleet's assistant and eventual replacement was Harold Sebring who would later sit as a judge in the Nuremberg Trials.[28] Football was formative for the men who defeated the nazis.

5-Star rank was not limited to the Army, nor was the Army the only service turning football players into senior officers. Admiral William "Bull" Halsey served his time as a fullback for the Naval Academy at Annapolis and was president of the Academy's Athletic Association.[29] In the South Pacific Theater, Halsey would replace Vice Admiral Robert Ghormley, a former running back and teammate of Halsey at the Academy.

Anyone attending a University of Tennessee football home game will walk through Neyland Stadium's gates, named after Brigadier General Robert Reese Neyland. Neyland, a graduate of West Point whose career included serving as an aide to General Douglas MacArthur and later served in China, Burma, and India during the Second World War, coached the Tennessee football team beginning in 1925

Fig. 2.2 President ford playing football for Michigan. *Source* Encyclopedia Britannica

while simultaneously instructing R.O.T.C..[30] He would take up the head coaching job following World War II and tally a record of 173-31-12. The exceptional record included four national and seven conference championships. General Neyland was not a unique case as other big-name schools hired either active or former military officers to coach their teams. The University of Nebraska, for example, hired coach Biff Jones, an army major before the 1937 season. Coach Jones would lead his team to a Rose Bowl appearance at the conclusion of the 1940 season, which they lost to Stanford. Coach Jones returned to active duty following the U.S. entrance into the Second World WarFig. 2.2.

For the 38th President of the United States, Gerald Ford, football was central to his childhood and adolescence. While attending and playing football for the University of Michigan, Ford's team won three national championships. Upon graduating from the University of Michigan, President Ford declined contract offers from both the Green Bay Packers and the Detroit Lions.[31] Compared to any other president in their physical prime, Gerald Ford's athletic accomplishments on the gridiron ranks him as perhaps the best athlete to ever serve in the Oval Office.

More than presidents playing football in their early lives, the term "football" represents the awesome power of the Presidency. Indeed, the president is the ultimate decision-maker for the employment of nuclear weapons. As such, the briefcase containing the nuclear codes for the President to authorize a nuclear strike is known as "the nuclear football," or simply "the football," and always within arm's reach of the President.

Just as football influences American strategic leadership, the military has influenced some of the greatest football minds. Some of the greatest football coaches of all time began their coaching careers at our nation's military installations. George Halas, the first owner of the Chicago Bears and one of the founders of the National Football League (NFL) played for the Great Lakes service team during the First World War and earned the title of Rose Bowl Most Valuable Player in the 1919 New

Year's Day contest against the Mare Island Marines.[32] Halas returned to the Great Lakes training facility to coach their football as part of his service in the Second World War.[33] When Halas designed the Bears uniforms, he included stripes around the sleeve, an idea taken from the gold stripes naval officers wear on their cuffs which designate their ranks.[34] Football's heritage reaches back to each of the services.

In his book *War Football*, Chis Serb describes how World War I was an interesting time when football and the military evolved together. During the time of a global war and a global pandemic combined with a national draft, colleges (including the service academies) placed lesser emphasis on football. Indeed, many college and former college stars were drafted into uniform leading to the rise of non-academy football teams. The 1918 Rose Bowl pitted Mare Island Marines against Fort Lewis's 91st Division. Mare Island would return for the 1919 game against the Great Lakes Naval Training Station making them the first team to play in back-to-back Rose Bowls. The service teams across the nation offered the first-time players could participate following their college glory years, and this combined with the national exposure of the game paved the way for the rise of professional football.[35] In this sense, football and the American Military are forever joined at the hip.

Just as World War I turned war into a global event, it did the same for football. Following the November 11th, 1918 armistice, the American Expeditionary Force hosted a football tournament with the championship game played at the Velodrome in Paris.[36] In front of over fifteen thousand fans which included General Pershing, the commander of the American Expeditionary Force the 89th Division beat the 36th Division 14–6.[37] Championing for the tournament was Lieutenant Walter Camp, known as the founding father of American Football, who served as the Navy's athletic director. Camp's "daily dozen set-up" exercises were a central part of physical fitness training at naval training centers during the Great War.[38] By the end of World War I, over 150,000 sailors and a million soldiers participated in the sport.[39] Meanwhile, on the other side of the ocean, Pershing's contemporary, Army Chief of Staff Peyton March had his personal football history as a former fullback for Lafayette University way back in 1881.[40]

Walter Camp was also a founding father of what would become the National Collegiate Athletic Association (NCAA). The NCAA started as the Intercollegiate Athletic Association of the United States (IAAUS). Major Palmer Pierce from the U.S. military academy served as the organization's first president.[41] Pierce's military career was distinguished having deployed to Cuba and Puerto Rico in the Spanish-American War, to China during the Boxer Rebellion, to America's Southern Border to chase Pancho Villa, and culminated as the commander of the 27th Infantry Division in the First World War. In the latter two, he was a subordinate of General Pershing.

General Pershing not only led the American Expeditionary Force in World War I but was the commander chasing down Pancho Villa on the Mexican–American border shortly before the onset of the Great War. During this expedition, various National Guard units set up and played their own tournament. A young Dwight Eisenhower coached the 12th Division All-Stars in the tournament.[42] Indeed, football has a long history as a source of American military morale, welfare, and recreation (MWR).

Fritz Pollard, the NFL's first African American NFL head coach served in the U.S. Army during World War I. Before the war and in his college years he played for Brown University as the school's first black football player.[43] His talents helped the team play in the 1916 Rose Bowl. In 2005, he was inducted into the Professional Football Hall of Fame. Trailblazing ran in the Pollard family as Fritz's younger brother Frank Pollard was the first and only African American to play for a World War I service football team. By the end of World War I, approximately 150 NFL players would set aside their football uniforms in exchange for a military uniform.

Service teams continued in the inter-war period through the end of the Second World War. Throughout the roaring 20 s, Marine Corps General Smedley Butler, the commander of Quantico formed a base team that played to a respectable 75–14-5 record from 1921–1930.[44]World War II also saw the return of service football teams in the homeland and overseas. Training bases once again formed teams that competed against college teams. Indeed, in 1944 nine service teams placed in the final AP top 20 polls, including four in the top ten.[45] Add two more if you include the service academies at West Point and Annapolis.

Overseas, America once again exported football to Europe in the Second World War. In the North African and European Theater of War, Americans hosted a series of "bowl games" including the Arab Bowl in Oran, Algeria where an Army team defeated a Navy team 6–0 in front of 15–25 thousand spectators. Other games included the Riviera Bowl in Marseille, the Mustard Bowl in Dijon, and the Spaghetti Bowl in Italy where the Fifth Army defeated the Twelfth Air Force 20–0.[46] The Americans scheduled the Champaign Bowl in France on Christmas Day in 1944, only to be overcome by the events of the Battle of the Bulge.[47] The enemy always gets a vote.

Physical fitness was an essential part of getting soldiers ready to fight on the battlefields in Europe and in the Pacific throughout the Second World War. To help create physically fit soldiers, the U.S. Army called upon Colonel Ted Bank, a veteran of World War I and former Michigan starting quarterback in the early 1920s. Bank would serve as the chief of the athletic and recreation branch within the army's special services division.[48] In this role, Bank pressed to have each company deploy with athletic kits which included footballs, basketballs, and boxing gear.[49] Football was a key part of fitness and morale for deployed soldiers.

The legendary coach Paul Brown spent his time in the military coaching the Great Lakes Naval Training Station football team to a 9-2-1 record in 1944.[50] Part of Coach Brown's staff was a young assistant coach named Weeb Ewbank. Ewbank was the head coach of the Baltimore Colts' 1958 and 1959 NFL Championship teams and later was the head coach who guided the New York Jets to victory in Super Bowl III. Vince Lombardi, the namesake of the Super Bowl Trophy, began his college coaching career at the United States Military Academy at West Point based on a recommendation of a member of the academy's coaching staff, Sid Gillman.[51] Lombardi was an assistant to Red Blaik who graduated from West Point in 1919. When Blaik was hired to coach the team, he was appointed as the athletic director and given the rank of colonel.[52] Sid Gillman served in the U.S. Coast Guard during

World War II, and today Sid Gillman remains the only coach elected to the college and professional football hall of fames.

Vince Lombardi's time at West Point and his relationship with Red Blaik were formative years of Lombardi's coaching time before becoming the offensive coordinator of the New York Giants (with Tom Landry serving as offensive coordinator) and later as head coach of the dynastic Packers of the 1960s. Blaik would send the young Lombardi to New York City to show and narrate West Point game films to retired General of the Army Douglas MacArthur at the Waldorf-Astoria Towers in New York City.[53] There is an indirect line from MacArthur's well-known quote, "There is no substitute for victory" to Lombardi's famous quip, "Winning isn't everything; it's the only thing."[54]

Lombardi's tenure as head coach of the Green Bay Packers centers on the multiple NFL and Super Bowl Championships his teams earned. In his biography of Vince Lombardi *When Pride Still Mattered*, David Maraniss describes the epic 1961 season and the run to Lombardi's first championship. During the season, three of his key players Paul Hornung, Ray Nitschke, and Boyd Dowler were activated in a call up of the Army Reserves. These three players were not alone, as more than two dozen players were a part of the broad activation of service members. Maraniss details how Lombardi used his military connections he built while coaching at West Point to ensure his players could compete in the critical games to include the NFL Championship game.[55]

It was Sid Gillman who gave the legendary Al Davis his first big break by hiring him onto the staff of the Los Angeles Charges. Perhaps one of the more passionate and complex football coaches ever, Davis would serve as the commissioner of the American Football League (AFL) and later owner of the Oakland/Las Angeles Raiders. Al Davis spent his early years as head coach of the U.S. Army's Fort Belvoir football team followed by a stint as a coach for The Citadel in Charleston, South Carolina.[56] Al Davis's impact on the NFL is long lasting, not just in his innovations of on the field tactics, but in the historic hirings of Tom Flores, Art Shell, and Amy Trask. The former two are the first Latino and African American head coaches in the modern era, the latter being the first female Chief Executive Officer (CEO) for a NFL franchise (Fig. 2.3).

Vince Lombardi coached the Green Bay Packers to multiple NFL championships and led his team to victory in the first two Super Bowls. The first four Super Bowls took place before the unification of the National Football League (NFL) and the American Football League (AFL). For over a decade, the AFL was the rival league to the NFL. To enhance the credibility of the AFL, league owners chose Joe Foss to serve as commissioner of the league. Foss served in the Marine Corps during World War II, earning the Medal of Honor for actions in the battle of Guadalcanal.[57]

Following the Packers dynasty in the 1960s, two teams competed for dominance in the 1970s, the Pittsburgh Steelers and the Dallas Cowboys. Tom Landry, head coach of the Super Bowl VI and Super Bowl XII Champion Cowboys flew B-17 s over Europe during WWII, completing 30 combat missions and survived a plane crash in Belgium. Landry's service extended to his family, as his brother was killed

Fig. 2.3 Joe Foss, first AFL commissioner. *Source* Wikimedia commons

in a B-17 over the North Atlantic in 1942.[58] Before donning the famous fedora on the sidelines, Landry wore the silver bar of a first lieutenant over the skies of Europe.

From the Super Bowl Era, other famous coaches also served their time in the military. Don Coryell, former head coach of the San Diego Chargers (and implementer of the "Air Coryell Offense") was an Army paratrooper in the Second World War.[59] Bill Walsh, former head coach of the 1980's San Francisco 49ers dynasty, served two years as an enlisted soldier at Fort Ord.[60] Walsh's coaching rival Bill Parcels began his storied career as a linebacker coach and defensive coordinator at West Point and later served as head coach of the U.S. Air Force Academy. Parcells' protégé, Bill Belichick, was introduced to football by watching his father Steve Belichick scout opposing teams for the U.S. Naval Academy.[61] Steve Belichick, who wrote the foundational book on football scouting, served in the Navy during World War II.[62]

In Super Bowl XXV, standing on the opposing sidelines to Bill Parcells' New York Giants was Marv Levy, head coach of the Buffalo Bills. Levy served in the Army Air Corps (the predecessor to the U.S. Air Force) during World War II. Super Bowl XXV is perhaps remembered best for the game ending on Scott Norwood's missed field goal, but also as the game that took place during the air campaign phase of the 1991 Gulf War. Leading up to the game, there was a national level discussion about the appropriateness of playing the game with the nation at war. President George H.W. Bush scuttled the conversation when he reminded the nation that domestic life continues, including the Super Bowl.[63] Today, both Parcells and Levy are enshrined in the Professional Football Hall of Fame in Canton Ohio and Bill Belichick is sure to enter the hall in his first year of eligibility.

At the college level, before the rise of the Big-10, PAC-10, and SEC as the powerhouses of college football the military academies ruled the field. West Point won three consecutive national championships from 1944 to 1946. Both the Army and Navy Military Academies produced multiple Heisman Trophy winners. The award for the nation's best college player has been awarded to West Point's Felix Blanchard in 1945, Glenn Davis in 1946, and Pete Dawkins in 1958. The latter attained the rank of Brigadier General (1-Star) while earning a bronze Star for his service in Vietnam.[64] The Naval Academy produced two winners; Joe Bellino in 1960, and Roger Staubach in 1963.[65] We may not see an academy football player compete for the Heisman, or an academy play in the college football playoffs these days, but one cannot have an informed discussion on the history of college football without speaking to the contributions of service members and the academies.

The Naval Academy's ties to football go deeper. Credit is widely given to then mid-shipman Joseph Mason Reeves as the first football player to wear a helmet in a game. Following a series of head injuries in 1894, as a member of the Naval Academy's football team, the superintendent of the Academy ordered Reeves not to play. Reeves proved resourceful and determined to convince the superintendent otherwise. With the help of a local shoemaker, they designed a leather helmet.[66] With the helmet on, Reeves played a role in defeating Army 6–4.[67] Reeves who served as Commander in Chief of the U.S. Fleet retired in 1936 only to be recalled to serve his nation in 1940 in the lead up to WWII. He again retired from the Navy following the war as an Admiral having served as the Navy's Lend Lease Liaison Officer and as chairman of the munitions assignment committee. Nearly half a century later in 1941, George S. Patton, perhaps while reminiscing about his days as a back-up, designed and wore a uniform for tankers that included a golden football helmet in lieu of the cap or hat worn by other army combat branches.[68] Of course, helmets are an essential piece of equipment for both football and military uniforms today.

It was the Navy that got Daniel Eugene "Rudy" Ruettiger his start in life. Before attending Notre Dame and walking onto the football team, Yeoman Ruettiger enlisted in the Navy serving two years at sea.[69] It was the G.I. Bill that paved the way for his college education and his famous football journey that was memorialized in the 1993 film *Rudy*, starring former Goonie Sean Astin.

Rudy isn't the only famous football name associated with the military and Notre Dame. The legendary Lou Holtz whose college coaching career included leading the 1988 Notre Dame team to a national title learned his first lessons in leadership as an R.O.T.C. cadet at Kent State. He served in the period between the Korean and Vietnam Wars as an infantry officer assigned to Fort Knox Kentucky.[70] Today Lou Holtz still credits his army leadership development as a key driver for his coaching success.

Every December college football teams compete in games rife with military connections: the Lockheed Martin Armed Forces Bowl, the Military Bowl Presented by Northrop Grumman, the Independence Bowl, and the Liberty Bowl. The Army,

Navy, and Air Force Academies compete for the Commander in Chief Trophy, awarded to the best football team among the three. American Presidents have a long history of attending the annual Army-Navy game and even participating in the coin toss. Dating back to 1890, the annual Army-Navy game is perhaps one of the most popular college football games of the season and one of the oldest rivalries in sports. The game holds a special place on the calendar occurring a week after all other Division I conference teams complete their schedules.

It wasn't just military operations in combat where football analogies and phrases come into the light. As the U.S. Legislature held hearings considering a transition from a conscripted force to an all-volunteer force in the early 1970s, a speaker invoked Vince Lombardi to finalize the argument in favor of the all-volunteer force. In testimony to the House Armed Services Committee (HASC), the army's Special Assistant for the Modern Volunteer Army (SAMVA), General Forsythe testified, "If Vince Lombardi sent his offensive backfield off to cut the grass on the football stadium one day, and the next day the tackles and two ends were out selling tickets (these jobs are 'mission related') he wouldn't have much of a football team."[71] Forsythe's statement was an effort to convince Congress that the focus of soldiers should be the job of soldiering and that in an all-volunteer army, to retain the best talent, other jobs and tasks not related to warfighting skills should be contracted out. The underlying philosophy remains in the armed forces today and expanded during the era of continual deployments to Afghanistan and Iraq.

Following the terrorist attacks on September 11th, the National Football League initiated a "Salute to Service" month each November to "honor servicemen and women from all branches and their families."[72] Players and coaches walk the sidelines in military themed activewear while flyovers and oversized flags accompany the singing of the national anthem, often with a color guard of uniformed service members. The practice has not been without controversy. Investigative journalism revealed that the Department of Defense spent nearly seven million dollars paying various professional sports organizations, including football, hockey, and baseball for various patriotic displays.[73] While many considered this a waste of taxpayer money, and in some ways poor taste, there is the argument that the Department of Defense in some ways understands that sports fans make up a significant part of the recruiting population.

At an individual level, there is a history of professional athletes in the National Football League transitioning their football life to service in the U.S. military. Perhaps the most famous example is Pat Tillman, who gave up a promising career with the Arizona Cardinals to enlist as an Airborne Ranger. Tillman would make the ultimate sacrifice for his nation, eventually meeting his end because of friendly fire in Afghanistan.[74] Joe Cardona, a U.S. Naval Academy graduate and 5th-round draft pick earned three Super Bowl rings as the long-snapper for the New England Patriots while simultaneously serving in the Naval Reserves. Although he didn't play in the NFL, retired Colonel Michael Steele, known from his days as a ranger company commander in Somalia, was an offensive lineman at the University of Georgia.[75] He would earn a championship ring when the Bulldogs won the 1980

Fig. 2.4 Roger Staubach in Vietnam. Photo from wikimedia commons

national championship. His head coach at Georgia was Vince Dooley, a veteran of the United States Marine Corps.

There are more recent professional football players who served in the military before entering the NFL. Chad Hennings, a 1988 graduate of the U.S. Air Force Academy earned the Outland Trophy as the best interior lineman in the United States. Henning would be drafted by the Dallas Cowboys but served his time as an A-10 pilot before joining his team and earning three Super Bowl Rings with the Dallas Cowboys in the early 1990s.[76] Steve Russ graduated from the Air Force Academy in 1995 and was drafted in the 7th round by the Denver Broncos. Following two years of active service, Russ joined the Broncos just in time to win back-to-back Super Bowls in 1997 and 1998. Russ would move on to full time coaching at schools such as Syracuse, Wake Forrest, and the Air Force Academy. In 2020 following a stint as a linebacker coach for the Carolina Panthers, Russ joined the coaching staff of the Washington Commanders.[77] Mike Anderson, the NFL's 2000 rookie of the year served in the Marine Corps before his time on the Denver Broncos. Alejandro Villanueva played football at West Point, earned a commission in the infantry, earned a ranger tab, and earned a bronze star for valor while serving in the 75th Ranger Regiment. Villanueva then became an offensive lineman for the Pittsburgh Steelers and later as a starting tackle for the Baltimore RavensFig. 2.4

The transition from football player to national leader is not a one-way street. Other notable football players with a record of service include Hall of Fame and Super Bowl winning quarterback for the Dallas Cowboys Roger Staubach, a graduate of the U.S. Naval Academy at Annapolis who served a year overseas as a supply officer in Vietnam. Staubach maintained his roots to the U.S. military and in 2010 lit the ceremonial torch to kick off the first ever Warrior Games. Another Annapolis graduate Phil McConkey earned a championship ring playing for Bill Parcells' New

York Giants, catching a touchdown pass in their win over the Broncos in Super Bowl XX. Like Staubach, McConkey completed his mandatory tour in the Navy before playing in the NFL.

The Vietnam War had an impact on the National Football League, though not quite as large as the mobilizations of the First and Second World Wars. While Roger Staubach served his time in Vietnam as a naval officer, his on the field rival, Pittsburg Steelers running back Rocky Bleier served as an infantryman after he was drafted in the U.S. Army following his rookie season. Bleier would earn a Purple Heart after receiving both a bullet and shrapnel in his leg during a firefight with the enemy.[78] Bleier would recover from his wounds and play a key role in the Steelers dynasty era throughout the 1970s.

The only NFL star to lose his life in Vietnam was former Buffalo Bill Lieutenant Bob Kalsu. Lieutenant Kalsu lost his life to enemy mortars.[79] If the name "Kalsu" is familiar to Iraq War Veterans, it's because Forward Operating Base (FOB) Kalsu located in Iskandariya was named in his honor.

Football is deeply intertwined with American culture. Football nurtured political and military leaders in their formative years, framing how they thought about life and conflict. In many cases, football teaches leadership, cultivates a strong work ethic, and instills discipline that is beneficial for both sports and military service. The attention to detail and anticipating adversaries encourage critical and strategic thinking. The leadership, work ethic, discipline, and attention to detail lessons learned practicing and playing the game apply to how leaders make policy and strategy decisions in the contemporary strategic environment.

Notes

1. Adam Lazarus, *Super Bowl Monday: From the Persian Gulf to the Shores of West Florida—New York Giants, the Buffalo Bills, and Super Bowl XXV*. (New York, NY: Taylor Trade Publishing, 2011).
2. Matthew Sherry, *Any Given Sunday: The NFL's Epic 100-Year History in 20 Games*. (London, UK: Orion, 2020), Kindle, Loc 5947.
3. Ibid.
4. Perhaps the best example of football analogies was performed by the late comedian George Carlin. The YouTube video can be found here: https://www.youtube.com/watch?v=aIkqNiBASfI
5. Operation Goal Post was the landing of General George S. Patton's Task Force on the coast of Morocco in the larger Operation Torch in World War Two.
6. An Allied airborne operation to cross the Rhine River in March of 1945.
7. Large scale bombing campaigns over North Vietnam in 1972.
8. Michael Holly. *War Room: The Legacy of Bill Belichick and the Art of Building the Perfect Team*. (New York, NY: Harper Collins Publishers, 2011).
9. John S. Watterson, "Political Football: Theodore Roosevelt, Woodrow Wilson, and the Gridiron Reform Movement." *Presidential Studies Quarterly*, 25.3 (1995). 555.
10. Presidents Who Played Football." Retrieved from: http://www.profootballhof.com/presidents-who-played-football/

11. Jeff Pearlman, *Football for a Buck: The Crazy Rise and Crazier Demise of the USFL*, (New York, NY: Mariner Books, 2019). Not only did President Trump own a team in the USFL, but he was also largely responsible for its demise.
12. Brian Kilmeade, *The Games Do Count: America's Best and Brightest on the Power of Sports.* (New York, NY: Harper Collins Publishers, 2004), 156–158.
13. John S. Watterson, "Political Football: Theodore Roosevelt, Woodrow Wilson, and the Gridiron Reform Movement." *Presidential Studies Quarterly*, 25.3 (1995). 556.
14. John J. Miller, *The Big Scrum: How Teddy Roosevelt Saved Football.* (Harper Collins, 2011), 14.
15. Jeremy Fuchs, "What If … Theodore Roosevelt's Son Hadn't Altered Football Forever?" *Sports Illustrated.* Retrieved from: https://www.si.com/specials/greatest-sports-what-ifs/college-football/2017/theodore-roosevelt-jr-harvard-football
16. Miller, 135.
17. Steven Norris, "Remembering Georgia Tech's Heroes." 26 May, 2016. Retrieved from: https://www.news.gatech.edu/2016/05/26/remembering-georgia-techs-heroes
18. Miller, 58.
19. John S. Watterson, "Political Football: Theodore Roosevelt, Woodrow Wilson, and the Gridiron Reform Movement." *Presidential Studies Quarterly*, 25.3 (1995). 55.
20. Lars Anderson, *Carlisle vs. Army: Jim Thorpe, Dwight Eisenhower, Pop Warner, and the Forgotten Story of Football's Greatest Battle,* (New York, NY: Random House Publishing, 2008).
21. Carlo D'Este, *Eisenhower: A Soldier's Life*, (New York, NY: Henry Holt, 2002).
22. Timothy P. Brown, *Fields of Friendly Strife: The Doughboys and Sailors of the WWI Rose Bowls.* (West Bloomfield, MI: Brown House Publishing, 2017). Kindle, Chapter 1.
23. Timothy P. Brown, *Fields of Friendly Strife: The Doughboys and Sailors of the WWI Rose Bowls.* (West Bloomfield, MI: Brown House Publishing, 2017). Kindle, Chapter 14.
24. Serb, 108.
25. Lars Anderson, *Carlisle vs. Army: Jim Thorpe, Dwight Eisenhower, Pop Warner, and the Forgotten Story of Football's Greatest Battle,* (New York: Random House Publishing, 2008).
26. National Football Foundation. Retrieved from: https://footballfoundation.org/sports/general/roster/gen--james-a---van-fleet/80
27. Norm Carlson, "Gator Talk—Strange Tales of Gator Coaches." Retrieved from: https://floridagators.com/news/2010/11/14/19205.aspx
28. Ibid.
29. "William Frederick Halsey, Jr. Naval History and Heritage Command. Retrieved from: https://www.history.navy.mil/research/library/research-guides/modern-biographical-files-ndl/modern-bios-h/halsey-william-f.html
30. David Vergun, "Sports Heroes Who Served: Tennessee Football Coach Served in World Wars I, II." DoD News. (17 November 2020) retrieved from: https://www.defense.gov/Explore/Features/Story/Article/2411661/sports-heroes-who-served-tennessee-football-coach-served-in-world-wars-i-ii/
31. Sarah Pruitt. "9 Things You May Not Know About Gerald Ford." 1 September 2018. Retrieved from: https://www.history.com/news/9-things-you-may-not-know-about-gerald-ford
32. Timothy P. Brown, *Fields of Friendly Strife: The Doughboys and Sailors of the WWI Rose Bowls.* (West Bloomfield, MI: Brown House Publishing, 2017). Kindle.
33. John Eisenberg, *The League: How Five Rivals Created the NFL and Launched a Sports Empire.* (New York: Basic Books, 2018), 13.
34. R.D. Rosen, *Tough Luck: Sid Luckman, Murder, Inc., and the Rise of the NFL.* (New York: Atlantic Monthly Press, 2019).
35. Brown, *Fields of Friendly Strife: The Doughboys and Sailors of the WWI Rose Bowls.* (West Bloomfield, MI: Brown House Publishing, 2017). Kindle.
36. Chris Serb, *War Football: World War I and the Birth of the NFL.* ((Lanham, MD: Rowman & Littlefield Inc, 2019), 10.

37. Russ Crawford, *Le Football: A History of American Football in France.* (Lincoln, NE: University of Nebraska Press, 2016). 23.
38. Whitfield B. East, A Historical Review and Analysis of Army Physical Readiness Training and Assessment. (Fort Leavenworth, KS: Combat Studies Institute Press, 2013), 57.
39. Serb, 34.
40. Serb, 162.
41. Brian M. Ingrassia, *The Rise of Gridiron University: Higher Education's Uneasy Alliance with Big-Time Football.* (Lawrence, KS. University Press of Kansas, 2012).
42. Serb, 117.
43. Amir Vera, "Fritz Pollard: First Black Coach in the NFL." Retrieved from: https://www.azfamily.com/community/black_history_month/fritz-pollard-first-black-coach-in-the-nfl/article_dff033b2-33e2-5049-a3f2-1a83d1e5eb06.html
44. Serb, 190.
45. Ibid.
46. Russ Crawford, *Le Football: A History of American Football in France.* (Lincoln, NE: University of Nebraska Press, 2016), 59–64.
47. Russ Crawford, xvii.
48. East, 87–88.
49. East, 91.
50. Doug Farrar, *The Genius of Desperation. The Schematic Innovations That Made the Modern NFL.* (Chicago, IL: Triumph Books, 2018). Kindle.
51. Timothy P. Brown, *How Football Became Football: 150 Years of the Game's Evolution.* (West Bloomfield, MI: Brown House Publishing, 2020). Kindle.
52. Lars Anderson, *The All Americans.* (New York, NY: St. Martin's Press, 2004), 66.
53. David Maraniss, *When Pride Still Mattered: A Life of Vince Lombardi.* (New York, NY: Simon and Schuster, 1999), 143.
54. The line is indirect. Although "Winning isn't everything, it's the only thing" is often attributed to Lombardi, and is indeed a phrase he often used and posted on locker room walls, as David Maraniss details in his book *When Pride Still Mattered*, the origin of the quote is from the movie *Trouble Along the Way*, starring John Wayne.
55. Maraniss, 284–289.
56. Mark Heisler, "Al Davis: the Machiavelli of Football." Los Angeles Times. 15 December, 1991. Retrieved from: https://www.latimes.com/sports/la-xpm-1991-12-15-la-sp-al-davis-heisler-story.html
57. Ian S. Kahanowitz, *NFL 1970: The Inaugural Season of the NFL.* (Sunbury Press: Mechanicsburg, PA: 2021), 25.
58. Mark Oristano, "How Legendary NFL Coach Honed his Instincts in World War II." *HISTORYNET*. 13 February 2022. Retrieved from: https://www.historynet.com/how-legendary-nfl-coach-tom-landry-honed-his-fighting-instincts-in-world-war-ii/
59. Bob Glauber, *Guts and Genius: The Story of Three Unlikely Coaches Who Came to Dominate the NFL in the '80 s*, (New York, NY: Grand Central Publishing, 2018). Kindle. Chapter 1.
60. Ibid.
61. David Halberstam, *The Education of a Coach.* (New York, NY: Hachette Books, 2012).
62. Steve Belichick, *Football Scouting Methods.* (Mansfield, CT: Martino Publishing, 2011).
63. Adam Lazarus, *Super Bowl Monday: From the Persian Gulf to the Shores of West Florida—The New York Giants, the Buffalo Bills, and Super Bowl XXV*. (New York, NY: Taylor Trade Publishing, 2011).
64. Chris Huston, "A Tribute to Heisman Veterans," (2019). Accessed 1 July 2020. Retrieved from: https://www.heisman.com/articles/a-tribute-to-heisman-veterans-2/
65. "The Heisman Trophy Winners List" Accessed 1 July 2020. Retrieved from: https://www.heisman.com/heisman-winners/
66. Thomas Wildenberg, *All the Factors of Victory: Adm. Joseph Mason Reeves and the Origins of Carrier Airpower*. (Annapolis, MD: Naval Institute Press, 2003), Pg.23.
67. Ibid.

68. Edward M. Coffman, *The Regulars: The American Army, 1898–1941.* (Cambridge, MA: Belknap Press, 2007). Kindle, Chapter 10.
69. David Vergun, "Sports Heroes Who Served: Navy Enlistment Helped Spark Football Legend's Career." DoD News. (12 January 2021). Retrieved from: https://www.defense.gov/Explore/Features/Story/Article/2463420/sports-heroes-who-served-navy-enlistment-helped-spark-football-legends-career/
70. John Raughter, "College football great Lou Holtz reflects on being named Legion's 2019 "Good Guy" " The American Legion. (23 October, 2019). Retrieved from: https://www.legion.org/convention/246961/college-football-great-lou-holtz-reflects-being-named-legion%E2%80%99s-2019-%E2%80%98good-guy%E2%80%99
71. Beth L. Bailey. *America's Army: Making the All-Volunteer Force* (Cambridge, MA: Bellcamp Press, 2009). Kindle.
72. NFLCommunications.com.
73. Eyder Peralta, "Pentagon Paid Sports Teams Millions For "Paid Patriotism" Events." National Public Radio. 5 November, 2015. Retrieved from: https://www.npr.org/sections/thetwo-way/2015/11/05/454834662/pentagon-paid-sports-teams-millions-for-paid-patriotism-events
74. According to NFL Communications, the NFL Salute to Service partners with the Pat Tillman Foundation among other charities as part of the Salute to Service.
75. This episode was captured in the book *Blackhawk Down* by Mark Bowden, which was then turned into a film by Ridley Scott.
76. Jill Lieber, "A New Blue Yonder: Former Air Force Pilot Chad Hennings Reports for Duty with the Cowboys." *Sports Illustrated.* 13 July 1992. Retrieved from: https://vault.si.com/vault/1992/07/13/a-new-blue-yonder-former-air-force-pilot-chad-hennings-reports-for-duty-with-the-cowboys
77. https://www.commanders.com/team/coaches-roster/steve-russ
78. Sherry, Chapter 10.
79. Jesse Berret, *Pigskin Nation: How the NFL Remade American Politics.* (Springfield, IL: University of Illinois Press, 2018), 71.

Chapter 3
International Relations Theory

> In the end, peace can be achieved only by hegemony or by balance of power. – Henry Kissinger.
>
> There is a way to win every game. Finding it is what separates some people from others. -Bill Parcels.

Before Before jumping into military theory and strategy, it is important to understand the various international relations theories that govern interactions between state and non-state actors. How one views the world tends to dictate how one approaches solutions to global problems. Each theory helps to explain why, or the purpose that political decision makers choose to employ the military instrument of national power. Understanding the foundations of theories such as realism, liberalism, and constructivism helps to frame discussions. It is important to note, that when national level leaders make decisions, they often sway back and forth between various theories. No single theory of international relations and power can account for why nations and states act as they do.

Because national level leaders move between the fundamentals of realism, liberalism, and constructivism, there are a host of other sub-theories that are useful in explaining state behavior. These theories vary from balance of power to collective defense and security, to cosmopolitism, and democratic peace theory. Each has a role in helping to understand international relations, and to an extent why nations go to war. This chapter will explain these theories while providing the relevant football analogies and how they may apply on the gridiron.

3.1 Realism

Realism is perhaps the best place to start in understanding international relations. At its core, realism is a pragmatic way of understanding power. Realism's premise relies on the philosopher Hobbes, who posits that there is a constant struggle for power among states. Further, realism sees military power as the defining feature of this power struggle.[1] Through the realist lens states are unitary actors that are aggressive,

D. Sukman, *American Football and the American Way of War*,
https://doi.org/10.1007/978-3-031-55345-5_3

and always seeking greater security and greater power.[2] This leads to the outcome of each state having responsibility for its own defense as it can only rely on itself for protection.

If each state is responsible for its individual well-being and security, then the international system is state based. There is no higher authority to govern or judge the actions or to control the behavior of states in this system. With each state acting on its own, it must rely on itself for protection and to advance what it deems as national interests. While legal principles may govern actions within the state, there is no legal principle to govern actions between states.

Through the lens of football, from the Pop Warner through the National Football League, realism is represented through the competition that occurs throughout the season. It is important to recognize that football is a sport organized by teams where each team manages its own internal affairs. Each team competes on its own, hoping to win each week to advance its place in the standings at the expense of other teams. A win in one team's column is a loss for another. Power is gained and lost on the field and no single team oversees this chaotic environment.

From youth to professional football, teams align themselves in divisions and conferences. Most often this is done to limit the length of travel. Travel across the country for college and professional teams, and state and local travel for high school and youth teams. Long flights and bus rides as well as changing time zones can wear down a visiting team providing a competitive advantage for the home team. Of course, the breakdown of teams into these divisions serves to create intense rivalries between the teams, and the schools, towns, and cities they represent. Just as teams that are close in geographic proximity have intense rivalries, so do nations and states that border one another. The German—Franco rivalry that lasted for a couple of centuries is one example, as are the modern-day India—Pakistan and China—Japan, and Saudi—Iranian relationships.

Notwithstanding current day alliances such as NATO, most conflict occurs between neighboring states. Recent wars between Armenia and Azerbaijan and between Russia and Ukraine highlight this tendency. Today America's military has a global reach leading to wars fought in distant and far-away lands. However, from the American Revolution and the War of 1812 through the Mexican–American War, the Civil War, and Spanish-American War, the historical norm before World War II was fighting adversaries close to or inside our own territory.

3.2 Liberalism

Competing with the realist worldview is liberalism. Liberalism is the belief that interdependence and liberal institutions will move the international system from one based on power, to one based on agreed rules and norms. Liberalism favors human rights, civil liberties, and government by consent of the governed. Further, rather than a continual competition for power within realism, liberalism recognizes that states are interdependent, and international organizations such as the United Nations, the International Monetary Fund, and the World Bank are central to international relations and governance.

The end of the Second World War, and later the end of the Cold War brought into the geopolitical environment new eras of international cooperation. Nations around the globe have come to recognize the value of a rules based international order, and the economic and security benefits that this order provides. Organizations such as the United Nations, the North Atlantic Treaty Organization (NATO), and its adversary alliance the Warsaw Pact brought together nations seeking to advance self and mutual interests. Friendly and adversary nations also came together and recognized broad based rules of the international order such as the United Nations Convention on the Law of the Sea (UNCLOS). Intergovernmental organizations such as the International Monetary Fund (IMF), the World Trade Organization (WTO), and the European Union (EU). Not only did states come together, but a wide variety of non-governmental organizations formed to cooperate and manage the international environment such as Doctors Without Borders, the World Food Program, and the Bangladesh Rural Advancement Committee (BRAC), the latter of which is the largest NGO in the world (Fig. 3.1).[3]

Football from the liberalist vantage point occurs off the field. While teams compete against all others on a weekly basis, football is an organized sport where teams array themselves into divisions and leagues based on conditions such as geography and the size of a school's student population. Below the professional level, league officials understand that schools and towns with large populations have more talent to choose from, and thus hold a competitive advantage over the schools and localities with smaller populations.

Fig. 3.1 NATO is a key organization in how America applies military power to solve global problems. Photo from: Wikimedia Commons

Further, teams meet on the field under agreed upon rules. A touchdown or a field goal is worth the same number of points for either team. Referees patrol the field and toss yellow flags to penalize a team when a player commits an infraction such as holding, pass interference, or roughing the passer. A ball kicked through the endzone results in a touchback, a defensive player crossing the line of scrimmage before the snap of the ball is a 5-yard penalty for offsides. The distance between goal lines is 100 yards, and 53 yards separated the sidelines of opposing teams.[4] All teams agree to these concepts.

More than agreement on the rules of the game, various leagues cooperate at the higher level of the game to maximize revenue and profit. In both college and the NFL, teams have revenue sharing agreements to ensure viability. Further, within the NFL, players and owners abide by agreements worked through the NFL Players Association (NFLPA). These agreements set limits to rookie contracts, govern the size of the roster, as well as set punishments for individuals who violate league policies. Although there is a history of contentious relations between owners and the NFLPA which includes a combination of player strikes and owner lockouts, a generally copasetic relationship between the two has led to an immense growth in league revenue. This in turn has drastically increased the value of every team in the league to the owners' benefit along with increases in players' salaries, from the top to the bottom of the roster.

Simply put, without the cooperation of owners, coaches, teams, and players, the sport would cease to exist.

3.3 Constructivism

Constructivism is perhaps the most difficult of the three main international relations theories to comprehend. Broadly, constructivism recognizes that the international system is what individual states choose to make of it. States' and international actors' choices will change depending on the situation and the consequences. In the application of constructivism, one must understand the totality of the motives, assumptions, and intellect of individuals who make decisions vis-à-vis war and peace.[5] The value of constructivism is that it recognizes the dynamic environment of geopolitics, and that leaders, nations, and organizations tend to worry about what their interests are in the moment and are less concerned about what interests are enduring.

In football, what each team and what each respective league values can change depending on each situation. Violence and hard crunching hits were once an important value to football players and the NFL as a selling point of entertainment. However, this value changes as the league, players, and fans gradually came to recognize that having key players injured takes away from the competitiveness and entertainment value of the sport. A second example of constructivism in football is how football coaches judge the decision making of their opponents. Coaches must use historical mindedness and look to past decisions of opposing players and coaches, and the situations in which they made those decisions to help predict what they will do in the future.

Just as there is no unified theory that can unite the fundamental forces of nature, there is no single theory that can adequately explain all the decisions and strategies of nations and their respective leaders.[6] In addition to the big three explained above, other theories of international relations come into play.

3.4 Balance of Power

The balance of power theory of international relations posits that attempts by one state or actor to gain power or conquer another region or state will be met with a counterbalancing action from one or more other states. Stability is the result of the latter counterbalancing action or the threat thereof. For centuries, statesmen have understood this dynamic and used it to create systems such as the "Concert of Europe" following the defeat of Napoleon at Waterloo. This balance of power between European states and alliances largely kept the peace for a century, falling apart with the First World War. Moreover, the Cold War between the United States and the Soviet Union relied on a balance of power, specifically regarding nuclear weapon stockpiles. The threat of mutual destruction prevented a direct military conflict between the two nations, limiting their military competition to fights between proxies in places such as Vietnam and Afghanistan. Indeed, one nation often stood down in situations when the odds of direct military conflict rose such as the Berlin Blockade and the Cuban Missile Crisis.

At levels below professional football leagues maintain a balance of power by organizing into divisions and levels. For example, in college football, based upon a school's size, and number of athletic scholarships, a team might play in Division I, Division II, or Division III levels. Within DI, there are A and AA levels that further classify who the major power teams are and who they should compete against. A breakdown in these levels (which also occurs at the high school level) ensures a competitive atmosphere for players and fans alike. These levels also help to prevent injuries by having similar physical sized matchups.

At the professional level, the balance of power theory applies in the National Football League in three ways. First through the annual draft, second by way of the salary cap, and by the rules of free agency. How NFL teams draft players is in order of worst record through Super Bowl Champion of the previous season. The underlying theory is that the worst team can add the best players to their roster to make them more competitive the next season. Indeed, teams often go through cycles of playoff runs followed by down years of losing records. These down years, often referred to as "rebuilding years" are an essential ingredient to the success of the league.

The salary cap is a rule that limits the amount of money a team owner can spend on their cumulative player's salary each season. The pie for each team is set, and every player's contract, big or small, takes away a piece of the pie. In this manner the wealthiest owners, or teams that make the most revenue cannot simply outspend other teams and sign the top players at each position. Every team has a range of talent, and each team must make decisions on what positions and which players to invest their money in. Other limiting factors that impact the talent pool for teams at each level include limiting the size of a roster and limiting the number of football scholarships a university can award. At each level of football players, administrators, and coaches understand that the game is better when the game is competitive.

When the NFL labor rules changed in 1993 to allow players to sign with other teams at the conclusion of their contracts, the free agency era replaced the era of football dynasties which largely came to an end.[7] Key players at every position could enter the marketplace and sign with the team of their choice (mostly based on salary bids). As players excel in their respective positions and win championships on one team, they are free to take their talents at a higher price to another team. With players shifting teams throughout their careers, one team cannot retain all the top tier talent at each position (the salary cap essentially prevents teams from signing all the best available free agents).

Winning at the highest level of competition is hard. Most games are competitive, and free agency, the salary cap, and the annual draft help to maintain a relative balance of power, ensuring one team cannot maintain dominance over the rest of the league over an extended period. Apart from the 2001–2019 New England Patriots, the system has held. Free agency finally broke the Patriots in 2020 with the departure of Tom Brady who left for the sunny skies of Tampa, Florida bringing a championship to the Sunshine State.

3.5 Security Dilemmas

In international relations theory, a security dilemma is when one nation increases its military strength, forms new alliances, or takes other actions meant to enhance its security. Other states will see this as a threat. The increase in security or strength is quickly matched by another nation to ensure they are not placed into a position of relative weakness. It is "keeping up with the Joneses," only the consequences can lead to the destruction of humanity. Understanding the dynamics of a security dilemma is important to aspiring strategists as they can worsen international relations with other powers by setting their own security requirements too high.[8]

There are multiple historical examples of security dilemmas. Perhaps most recent and well known is the relationship between the United States and the Soviet Union throughout the Cold War. Each nation racked up its nuclear weapons stockpiles while in fear of the adversary gaining a numerical advantage. This dilemma occurred with conventional forces too, as both the United States and the Soviets constructed large armies and navies. Other historical examples include the British and Germans with each respective web of alliances in the lead up to the First World War and the naval arms race between the United States and Japan prior to the Second World War. In the contemporary world there is growing angst that the security dilemma paradigm is developing in Asia, and more specifically in the South China Sea (SCS). As China continues to militarize the SCS and various artificial islands the United States and its allies such as Japan invest in and employ various capabilities.

In football, security dilemmas play out when one team develops or acquires talent at one position such as wide receiver, followed by a rival matching that talent with a cornerback. Teams use both the annual draft and free agency to acquire so-called weapons each year to gain a competitive advantage over other teams in the league. The acquisition of a powerful defensive end or linebacker is often matched by a rival's acquisition of a star offensive lineman. Other methods include devising new offensive or defensive systems to counter the accumulated talent of an archrival. This played out in the 1980s when Bill Walsh developed the West Coast Offense with its short and quick passing routes to counter Lawrence Taylor and the defense of the New York Giants. Indeed, teams drafting or acquiring specific talent to counter a rival team's strengths is foundational in seeking a championship run.

3.6 Hegemonic Stability Theory

This theory postulates that the international security system benefits from a single powerful and dominant state. This powerful state then enforces the rules and norms between other important, but not as powerful members of the international community. In the 19th and early 20th Centuries Great Britain served in this role and was supplanted by the United States following the Second World War. Particular to this theory is that the hegemon must have the capability to enforce the rules, the will to enforce the rules, and commit to the system that it enforces. Today, the rise of China as a potential regional or global hegemon is creating friction as although they may have the capability and the will to enforce rules and norms, they are fast to remind the global community that they did not have a seat at the table when the current rules and norms of the international rules-based order were written.

The football version of hegemonic stability theory is the emergence of dynasties. Each decade has seen one or two teams dominate the rest of the league creating a sense of stability and predictability. In the 1960s, this was the Green Bay Packers, in the 1970s it was the Pittsburgh Steelers and Dallas Cowboys. The 1980s saw Joe Montana and the San Francisco 49ers win 4 Super Bowl titles before the Cowboys retook their position as the dominant team of the 1990s with three Super Bowl wins in four years. For the first two decades of the 21st Century, the New England Patriots claimed this mantle. Each decade saw a few minor powers, such as the Vikings in the 70 s, the Giants, Bears, Redskins, and Raiders in the 80 s, the Buffalo Bills in the 90 s, and the Giants, Packers, and Steelers during the Patriot reign.

In 2020, the hegemonic era of the Patriots seemed to come to an end. The departure of Tom Brady combined with the effects of COVID-19 on the league brought an end to the Patriots' streaks of playoff appearances, Super Bowl appearances, and seasons with a winning record. However, with the victory of the Tampa Bay Buccaneers in the Super Bowl led by former Patriot quarterback Tom Brady, the theory expanded from hegemonic teams and coaches to hegemonic players.

3.7 Democratic Peace Theory

This theory is relatively easy to understand. It states, based on the historical evidence that democratic states are more peaceful than autocratic or monarchist nations. As such, democratic nations simply do not go to war against each other. The flaw to this theory however is that democratic nations still fight wars and can be just as likely to instigate or participate in a war as authoritarian regimes. The recent wars in Afghanistan, Iraq, and Libya show that democracies are not immune to war. Indeed, a common criticism of this theory is that it leads to violent behavior by democracies who search the world for authoritarian dragons to slay, making the world no less peaceful.

Democratic peace theory is possibly one of the better arguments for why realism is not universally applied. In his book, *The End of History and the Last Man*, Francis Fukuyama describes how true realists should advocate for an American invasion and takeover of Canada.[9] Democratic Peace Theory looks beyond interests and includes values, and more specifically how values impact government decisions. Ideology in Fukuyama's estimate matters when choosing allies, partners, enemies, and adversaries, and democratic allies tend to be stronger partners.[10] Values and ideals still matter.

It's difficult to find an apt comparison to football and suffice it to say I will not make one here. Football teams by their nature are non-democratic. Decisions at each level of the game are made by coaches, general managers, athletic directors, and owners, and do not come from a vote by the players or fans. However, fans vote with plastic cards in their wallets, and their decisions on whether to attend games or purchase apparel can weigh heavily upon their minds and influence how long coaches and general managers retain their employment by the owner.

3.8 Complexity and Systems Theory

While not strictly a domain of international relations, systems theory is profoundly useful in understanding relations between state and non-state actors and in how football games unfold. A system in the words of Donella Meadows, the author of *Thinking in Systems* is "an interconnected set of elements that is coherently organized in a way that achieves something."[11] This definition accurately describes the inner workings of military organizations, war, international relations, and football teams.

The contemporary strategic environment is not only composed of nation states, but of non-governmental organizations such as the World Food Program and Doctors without Borders, international organizations such as the United Nations and Interpol, and malign non-state actors such as Al-Qaeda and ISIS. What a systems perspective of international relations allows one to do is to account for and attempt to understand the collective picture that results from all the interactions between them. Indeed, not a moment goes by that a leader of a nation or other international actor isn't making a decision that will have an impact and require decisions from other actors.

Football teams must also account for all the sub-components that add up to the product on the field. There are players and coaches, but also referees, fans, general managers, team owners, and alumni organizations to name a few. However, there are also intangible elements that impact a football team and the game such as the weather and motivation or morale of the team that interact with the physical or tangible elements. The more elements within the football system that impact the outcomes or desired goals of each team and organization, the more complex the system. Eleven players on each side of the ball begin their interactions with every snap, and every interaction and decision can change the course of a game or a season. A missed block on one side of the formation may lead to a quarterback rushing a throw that results in an interception. A high snap on a field goal attempt may lead to an incorrect placement of the ball by the holder resulting in a last-second field goal attempt that sails wide of the goal posts. The outcome of every play is dependent on all eleven members of the team acting in concert.

Before coaches decide on the play call or the formation they will employ, they consider the talent levels of the players on the field and their respective physical condition (accounting for injuries). They may account for various weather conditions such as rain, wind, and even the location of the sun. Other factors include the game situation such as the score, the down, and field position at the time. Further, they consider the historical decisions of the other team's coaches based upon countless hours of watching and studying game films.

What is useful in applying a systems theory is that it allows decision makers to take big picture perspective of their environment while breaking it down into component parts.[12] This perspective allows leaders to decide where to intervene in the system to attain their desired goals. Military leaders use the systems approach to determine what parts of an enemy force to attack and what parts they can ignore and further consider the desired and undesired second and third-order effects of their actions.

3.9 Conclusion

The various theories of international relations rarely stand on their own. Indeed, attempts to apply a single theory to strategic decision making is a fool's errand and can lead to disastrous choices from placing value on consistency. Political and military leaders often pick and choose various aspects from each theory and apply it to their individual worldview. Understanding each broadens the options or approaches they can apply when making decisions or seeking to understand the contemporary environment.

Notes

1. Jack Snyder, "One World, Rival Theories," *Foreign Policy* no. 145 (Nov/Dec 2004): 52–62.
2. Nikolas Gvosdev, Jessica Blankshaim, and David Cooper, *Decision-Making in American Foreign Policy*, (Cambridge, MA: 2019), 66.
3. BRAC in Business, *The Economist* February 2010. https://www.economist.com/business/2010/02/18/brac-in-business.
4. This is the distance for American Football, Canadian Football fields are longer, and the rules governing pre-snap movement are different. Of course all Canadian football games follow the same rules north of the border.
5. Richard K. Betts, "Culture: Customs, Norms, and Learning." In *Conflict After the Cold War: Arguments on Causes of War and Peace*, 4th ed., edited by Richard K. Betts, (New York, NY: Routledge, 2016), 211–213.
6. 4 theories of physics: 1) general relativity or gravity 2) weak nuclear 3) strong nuclear and 4) electromagnetic.
7. Michael Oriard, *Brand NFL: Making and Selling America's Favorite Sport*. (Chapel Hill, NC: University of North Carolina Press, 2007), 175.
8. Robert Jervis, "Cooperation Under the Security Dilemma," In *Conflict After the Cold War: Arguments on Causes of War and Peace*, 4th ed., edited by Richard K. Betts. (New York: Routledge, 2016), 425–427.
9. Francis Fukuyama, *The End of History, and the Last Man*. (New York, NY: Simon & Schuster, 1992), 264.
10. Fukuyama, 280.
11. Donella H. Meadows, *Thinking in Systems*. (White River Junction, VT: Chelsea Green Publishing, 2008), 1.
12. John J. Tierney, "The Use Of Systems Theories In International Political Analysis." *World Affairs*, vol. 134, no. 4, 1972, pp. 306–324. *JSTOR*, www.jstor.org/stable/20671334. Accessed 7 June 2021.

Chapter 4
Strategic Theory

> Theory cannot equip the mind with formulas for solving problems, nor can it mark the narrow path on which the sole solution is supposed to lie by planting a hedge of principles on either side. But it can give the mind insight into the great mass of phenomena and of their relationships, then leave it free to rise into the higher realms of action. - Clausewitz.
>
> No plan of operations extends with certainty beyond the first encounter with the enemy's main strength. – Helmuth Von Molke.
>
> Every day you come into work, five things will happen you didn't anticipate – Bill Parcells.

Football is not only about the collision of force, but also the use of strategy, operations, and tactics. With a national culture centered on football, the jargon of the sport infiltrates everyday language. As football pervades and dominates American language and culture, it is useful to examine the theories of the great strategists, and the levels and principles of war through the lens of football.

Theory serves as a basic building block behind military operations from strategic planning to tactical execution. Understanding theory provides planners and operators with a framework to base their decisions. Indeed, learning theory will not teach someone what to think but rather provides ways to think. Multiple military theorists that span the range of warfare from the age of antiquity to the modern era, and many provide niche perspectives focusing on landpower, maritime power, or air power, respectively. Moreover, each theorist brings to the table a distinct cultural perspective. Clausewitz, for example, has a deeply Western perspective while Sun Tzu provides an Eastern view of strategy and war. As there is no one correct theory, theory will not offer specific answers to strategic problem sets, it will however with a broad understanding help the strategist or planner ask the right questions to develop options and solutions.

D. Sukman, *American Football and the American Way of War*,
https://doi.org/10.1007/978-3-031-55345-5_4

4.1 Clausewitz

Any serious discussion of strategy necessarily begins with Carl Von Clausewitz. Indeed, Clausewitz along with his Western contemporaries identify with a Western and American way of war. While Eastern concepts of warfare focus on defeating an adversary through deception and winning without fighting,[1] Clausewitz and Western theories of war recognize the inevitability of battle. Eventually, military forces will fight on the battlefield and influence the outcome of war to the advantage of one side. Football follows the Western way of war. While the Eastern way of war highlights winning without bloodshed as the highest form of skill, the Western way emphasizes violence. Teams or armies meet on the field to determine the outcome. This is the American way of war. In the words of former Secretary of Defense James Mattis, when discussing the war against ISIS, "we have already shifted from attrition tactics where we shove them from one position to another in Iraq and Syria, to annihilation tactics where we surround them" (Fig. 4.1).[2]

In his seminal work *On War*, Clausewitz compared war to a wrestling contest where two opponents grappled and maneuvered to get the other to submit. Had he been a 21st Century American, and not a 19th Century Prussian, Clausewitz would have probably made the football analogy in place of wrestling. The analogy works as in war, one does not plan or fight against an inanimate object. Rather, one plans

Fig. 4.1 Clausewitz. *Source* Wikimedia Commons

and fights against a living and thinking adversary. Clausewitz termed this reciprocal action.[3] Both military leaders and football players must constantly adapt to conditions on the battlefield and football field respectively. The team, or the military that learns the fastest holds a decisive advantage over their adversary.

Both armies and football teams enter war and the game expecting to win. Relying on the weakness, incompetence, or lack of preparation of an enemy team or force is a quick path to failure. No matter how well prepared a team may have been on the practice field, or in the gym, a competitor is doing the same thing, and sometimes with more commitment and devotion. Competitors plan to win, put in time at the gym, and adjust their game plan on every series. Success belongs to those who can stay one step ahead. Indeed, history is rife with examples of armies preparing for the last war instead of preparing for the next one.

During the years between the First and Second World Wars, some nations chose to develop a military approach based on the tactics, operations, and strategies of World War I. For example, the French built the Maginot Line, expecting to fight a static defensive war. Moreover, the French Army dispersed their tanks throughout their formations in support of their infantry. Juxtaposed against this thinking was the German Army, which built Armed (Panzer) divisions and corps, massing firepower, armor, and speed. The result of this divergent thinking was a six-week overwhelming victory by the German military and the rapid occupation of Paris in 1940.

However, at the strategic level, the Germans failed to learn their lessons from the First World War. Indeed, by 1941, the Germans were fighting a two-front war against a powerful alliance that held almost unlimited resources from a deeper and more efficient industrial base. Just as in the First World War, the Germans were unable to establish maritime supremacy over the British and Americans. The Nazi regime fought the Americans and British in the West and South, and the Soviets in the East. Moreover, the Germans were also fighting in the skies above their homeland. German operational warfighting prowess did not translate into strategic success.

In a similar vein, understanding that adversaries adapt leads to the wisdom of never underestimating a rival. A head coach may watch all the game films he wants, but there is another head coach on the other side-line developing a strategy of his own. Perhaps it is a trick play, or the exposure of a weakness on a team that never knew it existed. The history of the National Football League is replete with examples of underdogs defeating what was previously believed to be an unbeatable opponent. From the Jets beating the 18-point favored Baltimore Colts in Super Bowl III, to the Patriots defeating the 14-point favored St. Louis Rams and the Greatest Show on Turf in Super Bowl XXXIV. In both war and football, there is no such thing as a certain victory.

Identifying an opponent's source of strength and exploiting an opponent's weakness stems from the Clausewitzian concept of the center of gravity (COG), "the hub of all power and movement, on which everything depends, it is the point which all [the] energies should be directed."[4] The American military takes this teaching to heart. U.S. Joint Doctrine defines the center of gravity as "the source of power that provides moral or physical strength, freedom of action, or will to act."[5] American military planners spend countless hours analyzing and identifying enemy COGs.

Further, military planners devise strategies and plans to expose and attack what they believe is the enemy's center of gravity.[6] Witness the opening strike in Operation Iraqi Freedom in 2003, where the U.S. Air Force attempted (and failed) to kill Saddam Hussein on the opening night of the war. Planners believed Hussain was the Center of Gravity of the Iraqi Government. As the war evolved into an insurgency, the center of gravity switched to the will and support of the population.

The importance of the center of gravity is simple. Any side of a conflict has a set number of resources and capabilities they can apply to the problem. Identifying a center of gravity allows a military force or a football team to focus their efforts, and apply their respective strengths to attack an enemy, or to defend oneself. Proper planning identifies and analyzes both enemy and friendly centers of gravity, and in the case of more complex environments an analysis of other nations or non-state actors. These factors hold in both war and on the football field.

American football games and the game planning beforehand use the same center of gravity concepts found in U.S. military doctrine. Coaches and teams continually attempt to identify what an opposing team does best and take that capability away. Indeed, this concept is central to coaching great Bill Belichick's coaching philosophy.[7] This can be done through pressure on the quarterback or overloading an area of the field with defenders to take away parts of the run game. On offense, teams look to limit the opponent's best pass rushers through blocking schemes or expose weak pass coverage by continually passing in the direction of lesser talented cornerbacks and safeties.

Important to understanding the COG concept is the notion that centers of gravity can change or shift throughout a military operation. For example, the center of gravity at the onset of an operation may be the adversary's strategic leadership, an alliance, or the whole of their fielded forces. During the conflict, the regime may flee, the alliance may break, and the fielded forces may collapse. When the military operation enters a new post-conflict phase, the center of gravity may shift towards the will of the people or even shift to new and emerging leadership of a newly formed government. Identifying a center of gravity is a continual and evolving process. An easy example of a shifting center of gravity occurred in Iraq from 2003 to 2004. At the onset of the invasion, the U.S. identified Saddam Hussein and his inner circle as the Iraqi Center of Gravity. Once the regime fell, the COG shifted to the support of the population. This mirrored the change in the conflict from a major combat operation to a counterinsurgency fight.

When identifying a center of gravity, it is important to understand that there is no clear answer. Military analysts can spend countless hours examining the issue and identifying the center of gravity with its critical capabilities, requirements, and vulnerabilities only to get it completely wrong. War is uncertain, and one must always reexamine every step of the planning effort when things don't play out as anticipated. The best military leaders can see when their ideas of how to fight a battle or war are wrong and readjust their ways and means accordingly. In turn, the best coaches should understand when they have misidentified their own or opponents' strengths and weaknesses, or when those strengths and weaknesses shift throughout the game, through injury to a key player or other circumstances.

Clausewitz defined war and broke it down into its sub-units for the great captains of history. The old Prussian's thoughts on fog, friction, and chance are among the most useful takeaways from his magnum opus. Fog is the uncertainty in war, friction is the countless minor incidents that make the simple very difficult, and chance is the unpredictable circumstances that consistently occur in war.[8] This concept easily translates to the gridiron. The outcome of a football game and war are dependent on the interplay of skill and luck.

The fog in a football game is the uncertainty of an opposing team's play call, or when a quarterback is pressured during the chaos of blitzing defenders. We see friction when a head coach's headset loses communication with the quarterback, or when a referee makes a bad call. And we see chance when a 70-yard scoring play gets nullified due to a holding penalty on the opposite side of the field, wholly unrelated to the play. We see chance in who is sitting on the injured reserve list. Each game, indeed each play (not to mention off-field events) has the potential to eliminate the essential players of a team.

Clausewitz defined friction as the accumulation of seemingly minor things that can result in operational and strategic failure. For the military, these minor events can be a change in weather, the breakdown of critical weapons or transportation platforms, or the untimely illness of a senior leader. Friction is equally paramount in football. Dropped passes, missed blocks, blown coverages, and frivolous penalties can change the outcome of a game. Indeed, teams that lack discipline can cost themselves over 100 yards of field position leading to defeat at the end of the fourth quarter.

In 1944, the Allies executed Operation Market Garden. This was a bold offensive strike deep into German-occupied territory. Ultimately, the operation failed to achieve its desired objectives as the Allies were unable to overcome the accumulated friction over the course of the operation. Throughout the operation, the Allies were plagued by logistics shortages including not having enough transport aircraft. Paratroopers didn't always fall on the planned landing zone, and when they did, the landing zones proved to be too far away from their objectives. Bad weather and poor communications then made this movement more difficult than anticipated. Each point of friction by itself was not enough to doom the operation, but the cumulative effects proved impossible to overcome.

American football can bring in all three elements of fog, friction, and chance, often on the same play. In this sense, football offers a glimpse into complexity. A missed block or coverage assignment can lead to a sack and fumble or a 90-yard touchdown. On the other hand, a blown coverage may have zero impact on the game if the quarterback overthrows, or simply fails to recognize an open receiver. The outcome of every play and every game is dependent on the interaction of all 22 players on the field, and like war, uncertain in outcome.

While fog, friction, and chance are constant in war, there are ways to mitigate their impact. Training and experience, for example allow soldiers to build a tolerance for decision making under stress. Moreover, training under realistic conditions creates the space to develop contingency plans as various decision points come to fruition. Further, overwhelming mass, or size and firepower can sometimes make up for unexpected events in the heat of a battle. Finally, the ability of a leader to recognize the truth, and understand what is happening at a faster rate than his or her adversary thus maintaining the initiative, limiting options of the enemy, and dictating the tempo.

In addition to the concepts of fog, friction, and chance, Clausewitz presented his readers with the idea of Coup d'Oiel. Taken literally, Coup d Oiel is the French term for glimpse or glance. Within the context of war, a military leader can influence (but not direct) the probabilities of the outcome of a battle within the environment of fog, friction, and chance. Indeed, leaders with this trait immediately understand how a battle will unfold in front of them.[9] A leader who possesses the Coup d'Oiel sees, recognizes, and takes advantage of single or multiple opportunities presented on the battlefield that led to victory. Historic examples of military leaders who possessed the Coup d'Oiel include the likes of Fredrick the Great and Napoleon.

Every autumn weekend, America sees coaches, quarterbacks, linebackers, and cornerbacks who possess a Coup d'Oiel, and those who don't. Quarterback legends such as Tom Brady and Peyton Manning can use their vast experience to dissect opposing defenses as they walk to the line of scrimmage. Similarly, some coaches can spot weaknesses in an opposing team's offense or defense and concoct a scheme to attack said weakness. Moreover, the best quarterbacks gain the trust of their coaches and have the power to change the play call at the line of scrimmage. This authority does not come easy, rather it is dependent on a quarterback proving their abilities over time.

The opposite of a leader possessing Coup d'Oiel is incompetence. Military history does not have a dearth of incompetent leaders who led their forces to a disastrous end. Incompetence and errors befall every football team, both on the field and off. Interceptions, fumbles, missed coverages, and missed tackles are the errors of players, while bad clock management,[10] bad play calling, are the mistakes of coaches. Further, the incompetence of owners and general managers occurs through drafting of, or trading for, the wrong player.[11] In military parlance, these are examples of strategic miscalculation. Leadership in football and war is decisive to success and failure.

In warfare, incompetence is typically on display through the decision making of leaders. This may range from committing the strategic reserve at the wrong place or wrong time on the battlefield, to not recognizing the type of war a nation or force is fighting. The early part of the 2003–2011 war in Iraq put on full display how incompetence can prevent victory. Indeed, decisions such as the one to disband the Iraqi Security Forces shortly after the collapse of Saddam Hussain's regime led to a cohort of unemployed, vengeful, and trained adversaries to oppose the U.S. military occupation. Other historical examples of military incompetence include the British Charge of the Light Brigade in the Crimean War and the sheer ineptitude of military leadership throughout the various campaigns that led to the slaughter of millions in the First World War. For example, British Field Marshall Douglas Haig possessed a proclivity to try and turn any minor advance into a major breakout along the Western Front without any consideration for higher operational or strategic objectives. The commander of all British land forces was unable to think above the tactical level. Fortunately for the allies, this incompetence was matched by the German strategic leadership who also pursued tactical gains without deeper thought of strategic objectives.

The last aspect to understand from the writings of Clausewitz is that war's defining feature is violence. In chapter 1 of book 1 of On War, Clausewitz writes, "War therefore is an act of violence to compel our opponent to fulfill our will."[12] From the days of soldiers slashing each other with swords, to cannonballs ripping off heads and limbs, to the modern-day hellfire missiles turning humans into a pile of burning flesh, there is nothing glorious about the carnage of war. Indeed, Clausewitz warns his readers that without a true conception of the violence that accompanies war, one does not understand war.[13]

Of all the professional sports apart from boxing and ultimate fighting, football is the most violent of activities. While sports such as hockey and basketball characterize themselves as contact sports, football is a violent collision sport. Year in and year out images of football players being carted off the field with a serious knee, back, neck, or head injury are a common sight. The sidelines of NFL games hold an injury tent, where players receive immediate medical attention following action on the field.

The violence of both war and football can lead to life altering injuries. These injuries can be physical and visible in the form of a limp following multiple knee surgeries or they can be invisible brain injuries. While it is largely understood that the violence of war can lead to mental and moral injury in the form of traumatic brain injury (TBI) and post-traumatic stress disorder (PTSD), brain science is advancing to understand the impacts of repeated blows to the head that occur on the football field. Indeed, over the past decade multiple deaths of relatively young, retired athletes brought a focus on the dangers of playing football at all levels (Pop Warner, high school, college, and professional).[14]

While Clausewitz represents the Western ideals of war, it is the writings of Sun Tzu that exemplify the Eastern way of war. In *The Art of War*, Sun Tzu made multiple observations on how to fight a war. His theories complement Clausewitz's and apply on today's battlefields and football fields alike. These theories include the importance of deception, the necessity of a quick victory, an emphasis on knowing yourself and knowing an adversary, and the importance of psychologically defeating an adversary.

4.2 Sun Tzu

While Clausewitz epitomizes the Western way of thinking about war, it is Sun Tzu and his writing in *The Art of War* that embody the Eastern way of war. Knowing yourself and knowing your enemy is the first precept to understand to comprehend Sun Tzu's theory of war.[15] Any organization should understand their respective strengths and weaknesses. This includes the capabilities of the people within the organization. For example, military commanders must understand the strengths and weaknesses of their subordinate leaders, they must know the fighting strength, morale, and logistical situation of their forces before deciding to employ them in a fight. Conversely, leaders seek to understand the same conditions as their enemies and adversaries (Fig. 4.2).

Perhaps one of the most often quoted lines of Sun Tzu is that "all of warfare is based on deception."[16] From Odysseus's Trojan Horse of antiquity to modern day

Fig. 4.2 Procession of the Trojan Horse in Troy the Trojan Horse (*Source* By Giovanni Domenico Tipeolo. Image in the public domain from Wikimedia Commons)

cyber trojan horses, there are countless examples of militaries employing deception to trick their enemy throughout history. World War Two is filled with examples such as Patton's Ghost Army, largely made up of inflated tanks to convince the Germans of alternate D-Day landing sites. More recently the United States employed a feint before the ground invasion of Iraq in 1991 to occupy Saddam Hussain's army in Kuwait.

There are countless examples of deception in a football game. From the simple counter running play to play-action passes, to more complicated trick plays such as fake punts and fake field goals. Defensive coordinators today try to hide their coverage schemes before the offensive snap in their attempts to fool the quarterback into throwing the ball to the wrong place.

According to Sun Tzu, a key aspect of knowing yourself is understanding the talent you have and employing that talent in the proper manner. Indeed, Sun Tzu warns his readers to "not demand accomplishment of those who have no talent."[17] From the founding of the Prussian Kriegsacademy (War College), through the U.S. Army's talent management systems in 2020, aligning the right person to the right job at the right time remains one of the most consequential aspects of preparing a military for war.

From Pop Warner level football through the NFL, coaches continually employ their players in ways that set up the team for success. Often, football coaches do not have complete control or decision-making authority over the composition of the entire team. It is incumbent upon the coach to put players in a position to maximize their skills. A team whose players are small and fast ought to employ an offensive or defensive scheme that leverages the skill set.

There are other aspects of Sun Tzu that while not solely applicable to football, are important to understand. For example, Sun Tzu lists five fundamental factors of warfare which include (1) moral influence or passion, (2) weather, (3) terrain, (4) command or leadership, and 5) doctrine.[18] These fundamental factors are paramount in a coach's decisions on how to prepare and execute a game plan.

At the strategic level, weather, specifically the harsh Russian winter was decisive in the disintegration of Napoleon's Army, and again played a role in the defeat of the German Army in World War II. Further, weather often plays a role in the tactical and operational level decisions of warfighting commanders. Perhaps one of the more famous examples of weather influencing American military operations was the delay of the Normandy landings in June of 1944. Then General Eisenhower had to wait until the passing of a storm to initiate operations, ensuring that the amphibious force had air support. There was a short window of time to initiate the operation, and Eisenhower capitalized on it. Later in 1944, the German Army was able to expose a weakness in the Allied lines in the Ardennes. Their counterattack, now known as the Battle of the Bulge depended on cloud cover to prevent Allied air attacks against German ground forces.

In football, weather often plays a role in how a game will play out. Although teams play in the same conditions, coaches must base their decision making on the weather. For example, a game played in windy conditions will impact the coach's decision on fourth down on whether to punt, attempt a field goal, or go for the first down. Rain and snow may impact how well a quarterback can throw the ball, how a receiver can catch the ball, and how well a running back can handle the ball without fumbling. Further, the heat in Miami or the bitter cold of Green Bay can make a team unaccustomed to the conditions uncomfortable, providing a decisive advantage to the home team. Football teams often acquire personnel who can excel in the weather conditions of their home stadium. One of the intangibles of football players is how they perform in an enclosed stadium, or how they can adjust to cold, warm, or windy conditions.

In military strategy, it is critical to understand the terrain and geography at hand. At the strategic level, a nation's geography often drives the types of military capabilities that it develops. Further, geography will also determine the military strategies and plans a nation develops and employs. With regards to military capabilities, island nations such as Great Britain, and nations with large coastlines such as the United States have built powerful navies to achieve national interests. Moreover, nations such as Germany, which resides in the center of a continent, and Russia which lacks sufficient access to warm water ports have historically relied on large and powerful land forces to attain their strategic aims.

In football terms, the geography and terrain still matter. In geography, the location of a stadium and the game is often an essential ingredient in the outcome of the game. Games played at Denver's Mile High Stadium for example often impact the aerobic readiness and stamina of a visiting team. Further, teams often travel across time zones in away games, resulting in a disruption to a player's internal clock. While not significant to the average person, in the realm of professional football, this seemingly small change can be a decisive advantage to a team not traveling.

Command and leadership are foundational to the way every military organization performs in combat. The best strategies and plans can fail due to inept and incompetent leadership. While the best leadership might not win a battle or a war, bad leadership can lose a war. There is no shortage of examples of bad leadership in American military history. The recent wars in Iraq and Afghanistan offer some profound examples such as the prisoner abuses at Abu Ghraib in 2004.

Doctrine is the foundation upon which military organizations act. In the words of former Army Chief of Staff Gordan Sullivan, doctrine is "the Army's collective understanding of how we will fight; it's not what to think but how to think."[19] Doctrine provides a common language to military organizations and a common framework to operate from during times of crisis that require rapid decision-making. In some ways, doctrine serves as a heuristic or rule of thumb for leaders to follow.[20] Without doctrine, every unit from squads and platoons to divisions and corps would act in separate ways, creating chaos and friction between friendly units and forces. Further, without doctrine, military organizations would have no common set of tasks or battle drills to train on, setting conditions for failure in combat. A power of doctrine is that commanders will generally know what units to their left and right, and what higher and subordinate units will do without having to ask. A painful lesson in the lack of doctrine occurred in Iraq when at one point the U.S. Army had scores of battalions trying to conduct counterinsurgency (COIN) without a current COIN manual.

Just like various armies have their own doctrine, each football team has its own unique doctrine, also known as a playbook. Not only are players expected to memorize playbooks hundreds of pages long, but they are expected to recognize their opponents' formations, and quickly adjust their formations and play calling. Moreover, each team crafts its playbook to account for the talent level of the players on their team.

4.3 Thucydides

While military theorists and strategists may fall into a Clausewitz or Sun Tzu camp, there is general agreement on the importance and relevance of Thucydides. In his *History of the Peloponnesian War* Thucydides' writes a comprehensive history of the conflict which provides many of the same lessons we learn in modern conflict. From the value of alliances to judgments on strategic leadership through the concepts of conflict termination, the study of Thucydides is mandatory in the nation's war colleges and other professional military education institutions.

The first aspect of Thucydides and the Peloponnesian War that is remarkably like football is the warfare between various city states in ancient Greece. Athens and Sparta were the two most powerful states, and as such saw each other as primary adversaries. However, other city states had an impact on the war to include to include Corinth, Aegina, Thebes, Paros, Thasos, Samos, and Lesbos to name a few. In the modern era, football teams at each level represent towns and cities throughout America. Large cities such as New York, Chicago, Miami, Los Angeles, and Houston host professional teams, while college teams often represent both their school and state. At the youth level, football teams represent their small towns and high schools. Just as there was no Greek military in the time of Thucydides, there is no single American football team today. Football is a constant struggle of one city or locality against another.

If there is one concept in Thucydides writing that exemplifies football it is found in the Melian Dialogue. It is in the Melian Dialogue where we find the foundation of the expression "might makes right." In this section of Thucydides' writing, the Athenians approach the leaders of Melos and advise them to submit to Athenian rule, as they are too weak to contest the Athenians. Athens makes it clear to Melos that foreign domination is better than destruction. Indeed, the Athenian delegation presses Melos on the point that honor is a luxury of the strong with the phrase "the strong do what they will, the weak do what they must."[21] Today this dialogue is often used as an example of true realism, as discussed in the previous chapter.

Although football teams do not have the option to simply submit to a stronger team, generally the game is about power. Stronger teams, especially at levels below the NFL tend to dominate weaker teams. At the college and high school level, larger schools with more money and larger populations tend to field better teams than smaller schools with smaller budgets. The consistent appearance of large state schools such as Alabama, Ohio State, Florida State, University of Southern California (USC), Texas, and so forth in the college football playoffs with only an occasional appearance by a smaller school lends credence to this phenomenon.

A second lesson from the writing of Thucydides is that small nations and second-rate powers can play an oversized role in international relations. Small countries who act as proxies for large nations can force great powers to make strategic level decisions, and in the twenty-first century, these proxies can include non-state actors. America's NATO allies bordering Russia fall into the former category while Hezbollah, a non-state actor, serves as Iran's proxy in Lebanon and often forces Israel to make strategic decisions on the use of force. This aspect of warfare played out during the Cold War, when the United States became decisively involved in the Vietnamese Civil War, followed by Soviet military commitments in Afghanistan.

Each season, at all levels of football, weaker teams often look to play the spoiler against stronger teams. Smaller or weaker teams consistently play a role in football. Looking to play spoiler, small and weak teams often perform at their best when playing larger teams. Further, a win by an underdog or last place team can determine a strong team's seed in the playoffs, turn a first-place team into a wild-card team, and cost them home field advantage and a better shot at a championship. Further, not properly preparing to play a lesser or weaker team can lead to injuries to star players, again costing the stronger team a chance to make a championship run in the postseason. In seasons that range from ten to seventeen games, every game matters. Strong teams should never overlook what they think is a weak team on their schedule. Indeed, stronger teams should never decide to rest their starters and key players until after their playoff position has been determined.

Perhaps one of the more moving or emotional sections of Thucydides is the moment when Pericles gives his funeral oration. Pericles, the leader of the Athenian

Army uses the moment to remind the citizens of Athens of their values, of the greatness of their culture, and to appeal to their patriotism to honor those who have already perished in the war with Sparta. Pericles gives this speech at a time when the Athenians were growing tired of war and were failing to see the benefits of his strategy.[22] Ultimately, Pericles was giving an early version of a pre-game or halftime motivational speech.

Head coaches at all levels often give their team a motivational speech in the locker room before their team takes the field. While the overall talent and game plan are responsible for determining winners and losers, an effective motivation speech can fire up the emotions of a team and provide an advantage, however temporary, to one side or another. Of course, players from both teams are often at the receiving end of the speech, thus again limiting their overall value. Moreover, as players rise in age and level of the game, motivational speeches have less impact. While 10-year-old Pop-Warner and high school level players are influenced by the words of their coach, 28-year-old grown men who make millions of dollars don't need another adult revving them up for what is their trade and profession.

The expansion of war aims is one of the more important lessons from reading Thucydides and comes from his description of the Sicilian expedition. In this section of his book, Thucydides describes the decision of the Athenians to invade Sicily following a spirited debate among the city's leaders.[23] The arguments against the invasion ranged from Sicily serving as a distraction, to the long lines of communication between Athens and Sicily, and the need for Athens to consolidate and rebuild its strength. On the other hand, the arguments for the invasion included a need to assist an ally in need and to take advantage of an opportunity to strengthen the Athenian empire. The Athenians would lose and serve as a warning for future generations to think about war aims from the beginning of a military expedition and the consequence of changing the war aims during the conflict.

Athens did not possess sufficient power to adjust and expand its war aims in the middle of a war. Perhaps the perfect example of changing war aims is the United States in Iraq and Afghanistan from 2001 – 2021. In both wars, the immediate goals of overthrowing the Taliban and dismantling Al-Qaeda in Afghanistan and removing the Saddam Hussein Regime in Iraq occurred relatively quickly. Expanding the war aims to include the building of democracy didn't turn out so well. Success early in a war is not a guarantee of continued triumph.

As the football season moves from the opening day kickoff, coaches and general managers must continually reevaluate their season's objectives as the wins and losses pile up. High expectations at the start can quickly fade, forcing coaches to reconsider playing key players towards the end of the season, with the goal of preserving them for next year's run. General managers must consider the retainment of coaches and players should they fail to meet expectations. On the other side of the coin, teams that do unexpectedly well in the regular season may choose to make a run at a championship by trading for a key player before the trading deadline. This might mean sacrificing future draft choices or younger talent for an older prover player. War and competition are dynamic, and generals and coaches alike must continually evaluate their strategic positions to inform their future decisions.

4.4 Jomini

If there is one theorist whose ideas infiltrated the American way of war, it is Baron De Jomini. Jomini who fought on the other side of the battlefield from Clausewitz posited that there were specific principles or maxims that if followed properly could lead to victory. Indeed, while Clausewitz focused his writings on broad theory, Jomini's focus was on how to fight a war. American military leadership has followed his ideas from the Civil War through current service and joint doctrine often expressed as the elements of operational design (Fig. 4.3).

The key concepts of Jomini flow from the idea that one could break down warfare in a geometric manner. Lines of operation within a theater of war along with terrain focused decisive points were foundational to the development of a military strategy. Ultimately, Jomini posited that by maneuvering the mass of an army and throwing that mass at a decisive point, the enemy would be overwhelmed. Thus, it was up to the military commander to arrange his forces in a precise way to employ this mass. The obvious criticism of this idea is that war is uncertain, filled with the Clausewitzian ideas of fog, friction, and chance, and thus one cannot simply attempt to solve war through a series of maxims or mathematical equations.

On the gridiron, there is no shortage of coaches who believed that if their methods and playbooks were precisely followed, victory would surely follow. Of all the coaches that fit in Jomini's mold, perhaps the most significant is Bill Walsh. Walsh

Fig. 4.3 Antoine-Henri Jomini. *Source* By George Dawe Image in the public domain from Wikimedia Commons

trained his team to run precise routes and patterns as a part of his West Coast Offense. Precision was foundational to creating space between receivers and the defense.[24] Further, precise routes allowed the quarterback to throw the ball to a location, knowing the receiver would be there when the ball arrived. Like Jomini, the contest could be broken down to angles and movements on the field.

A second aspect of Jomini, and a key difference between him and Clausewitz is the emphasis on the attack. Where Clausewitz posited that the defensive was the strongest form of warfare, Jomini believed that the path to victory resides in attacking the enemy. Jomini's concept of the attack revolved around the maneuver of the mass of an army against fractions or portions of the enemy's force.[25]

Often, a football team is defined by how they perform on either offense or defense. Dominant defensive teams include the Minnesota Vikings Purple People Eaters of the 1970s, the Monsters of the Midway and the 1985 Chicago Bears, the 1986 New York Giants, the 2002 Baltimore Ravens, the 2013 Legion of Boom Seattle Seahawks, and the 2015 Denver Broncos. Each of these teams invested in star players on the defensive side of the ball, relying on the defense to carry them to the playoffs, and in the case of the Bears, Giants, Ravens, and Broncos, a Super Bowl championship.

In some cases, a team may invest in offensive talent, creating juggernauts who seemingly score at will. Throughout the late 1990s, the Buffalo Bills ruled the AFC with their K-Gun offense. In the mid-2000s, the New England Patriots worked themselves to a perfect 16–0 regular season via its unstoppable offense. However, neither the 1990s Buffalo Bills nor the 2007 New England Patriots would win the Super Bowl. Each would lose to a team with a strong defense, showing once again that the outcomes of both football and war will always remain uncertain.

In both football and war, there is no set rule or formula on which form of warfare will be successful. Often, victory requires a balanced effort. Leaders in both must decide where to apply their resources, often based on an assessment of their strengths and weaknesses, and those of the adversary.

4.5 B.H. Liddell Hart

One of the first strategic theorists to emerge in the twentieth century was B.H. Liddell Hart. Hart's strategic theory centered on what he identified as the indirect approach. Before reading Hart's work it is important to understand his background. Hart was a British officer whose experience on the Western Front shaped his way of thinking. The long-drawn-out battles of trench warfare, where attacking the enemy head on while charging through no-man's land, proved to be a fruitless endeavor. This lesson stuck with Hart in his search for a better way of war.

Hart defined strategy as the "art of distributing and applying military means to fulfill the ends of policy." In this sense, strategy was confined to the military instrument of national power. This however did not account for other instruments of national power a government has at its disposal. This leads concept of grand strategy. To Hart, there was the military strategy, which was confined to the movement and maneuver of military forces, and a higher-level grand strategy which accounted for other instruments of national power such as diplomacy, and economic considerations.

A coach is responsible for a team's strategy. The type of offensive and defensive formations a team employs, as well as the plays a team runs on every down, is the domain of the coach. Each week the head coach develops a theory of victory on how their team will defeat an opponent. Grand strategy is the domain of a team's owner in the NFL, the university at the college level, and the high school or local town at the youth level. Coaches may be responsible for wins and losses, but the owner is concerned with marketing, selling out stadiums, television contracts, and developing other sources of revenue.[26] Grand strategy in football is all encompassing and covers the activities of the entire organization.

4.6 Douhet

The next great strategist to examine is Giulio Douhet. Douhet advanced a theory of airpower that postulated the ability to gain command of the air would allow a force to break an adversary's will through the destruction of strategic vital centers.[27] Douhet believed airpower, unconstrained by geography, provided the opportunity for decisive offensive operations that technology had denied to ground forces. Predicating this belief is the idea that modern war (as Douhet saw it), was not a military matter alone, but rather between the people of opposing states. Applied to the contemporary football field, numerous offensive systems have developed over the years that rely on the passing game. Indeed, in today's game, it is passing and attacking opposing teams through the air that tends to be decisive in attaining victory.

The acceptance of Douhet's premise begins in World War II with strategic bombing. Both the Axis and the Allies used extensive air campaigns that included the bombing of cities to strike fear in the population. The London Blitz and the fire-bombing of Dresden serve as two stark examples. In the Pacific, the United States decimated Japan through a series of firebombings that destroyed every major city through the air, culminating with the dropping of atomic bombs on Hiroshima and Nagasaki in the summer of 1945. In more recent times, Operational Allied Force, the war against Serbia in 1999 was conducted by NATO through the air. NATO ground forces in this campaign operated on the periphery while attacks on Serbian forces and Serbian infrastructure came from the air.

The counter to this argument is that wars are not won through air campaigns alone. The fundamental problem with reliance on air power alone is that war is won or lost on the ground. Airpower cannot hold terrain. Douhet may have been appalled by the barbarity of trench warfare of World War I and hoped his theory would shorten conflicts and ultimately save blood and treasure, but to state the obvious, people live on land, and occasionally visit the air. The ability to control or deny control to an adversary is leveraged on the ground. While airpower may be decisive in an environment of total war, this condition rarely exists. In the end, airpower must be employed with other forms of combat power in the land, maritime, space, and cyberspace domains, and across the tactical, operational, and strategic levels of war.

Where Douhet and his airpower theory contemporaries are correct is that airpower can extend the operational reach of a military force.[28] Airplanes can reach far beyond enemy lines in ways that landpower cannot. Airpower and theory are also unique in that the air domain as a part of war has been around for a little over a century, a minute sample size considering the thousands of years and wars conducted in the land and maritime domains. Douhet's theories for example were not based on historical analysis and case studies, but rather through innovative thinking. In the modern era, we see the same paradigm when thinking about war in space and cyberspace.[29]

Today, the U.S. military operates as a joint force, applying combat power and capabilities in the air, land, maritime, space, and cyberspace domains. Similarly, the teams who most often enjoy victory are those who run a balanced attack; the ability to both run and pass the pigskin when and where they choose. The two greatest comebacks in NFL history came about from the failure to control the ground. Running the ball and keeping the play inbounds keeps the clock running, an advantage for a team with a large lead while incomplete passes stop the clock, providing the losing team time and space for a comeback. The Buffalo Bills overcame a 32-point deficit to defeat the Houston Oilers in 1993, in the largest playoff comeback at the time, and the New England Patriots coming back from a 25-point deficit to defeat the Atlanta Falcons in Super Bowl LI came about due to the inability of the Oilers and Falcons to control the ground game and run the clock in the second half. Airpower alone is not enough to win championships, teams must have a competent ground game, and a solid defense to balance the offensive attack.

There are numerous military examples of nations who tried to win wars through airpower alone. In World War II, the Germans attempted to force British capitulation through the London Blitz. This type of bombing campaign was repeated over the skies of Germany by the Allies. In either case, strategic bombing failed to attain a decisive victory. The famous general Curtis LeMay once remarked that the path to victory over the North Vietnamese was to "bomb them back into the stone age."[30] In 2003, the United States employed a "Shock and Awe" air campaign over the skies of Baghdad. While effective at the tactical and operational level, the Hussein regime was able to cling to power up until the point when U.S. ground forces captured the Iraqi leader. Further, airpower alone could not control the population, nor provide a safe and secure environment for the Iraqi people and government to flourish. In 2006, the Israeli Air Force hit every target on the Hezbollah target list, but Hezbollah's control of the Lebanese terrain, and subsequent messaging overcame Israeli firepower in the 34-day war.

4.7 John Boyd

The fighter pilot John Boyd is sometimes referred to as the greatest military thinker of the latter half of the 20th Century. Boyd, in his briefings, developed the concept of the OODA Loop. OODA—which stands for Observe, Orient, Decide, Act—is a method to get within an adversary decision cycle.[31] Although this idea was formulated for fighter pilots, the ability to paralyze an opponent's decision-making can lead to rapid success, both in air-to-air combat, as a method of military strategy, and on the gridiron.

The application of the OODA Loop was in thinking that the key to success was to maintain the initiative through faster decision-making. This fast or rapid (but still informed) decision-making can paralyze or freeze an adversary into inaction or poor decision making. In terms of military theory, the downside is that it assumes a competition between two sides when the American way of warfare is to fight alongside and with allies and partners. At the strategic level, slower decision making among allies can be a decisive advantage, as the consequences and risks of decisions at the strategic level are greater than those at the tactical level. Vladimir Putin may have a faster unilateral decision cycle than an American President who consults with allies and partners, but those decisions while leading to tactical success can have strategic drawbacks such as expulsion from the G-8, economic sanctions, the destruction of land forces, and international isolation, all of which occurred following Putin's decision to expand the invasion of Ukraine in early 2022.

The underpinning of Boyd's theory is that war has always been and will always be a human endeavor. Technology and weapons systems continue to advance, becoming more precise and more lethal, but to attain victory you must get into the mind of your enemy. A faster decision cycle was the way to do this.

To see Boyd's OODA Loop in action on the gridiron, witness how the advent of the no-huddle offense revolutionized offensive tactics in football.[32] Rapid offensive movement denied the ability of opposing defenses to observe the offense and decide how they should defend. The recent ending to Super Bowl XLIX serves as another example. Coach Belichick of the New England Patriots made his decision on the defense, refusing to worry about the time remaining, and subsequently putting all the pressure on Pete Carroll of the Seattle Seahawks.[33] The result was a 1-yard pass intercepted in the end zone while the league's best running looked on, helpless to do anything about it. The ability to get into an adversary OODA Loop is a decisive point where one force starts winning, and the other starts losing.

Perhaps the most visible marriage between football and Boyd's OODA Loop is pre-snap actions. Quarterbacks scan the defense and send players in motion to determine the defensive formation.[34] Simultaneously, the defense attempts to disguise its coverage. The quarterback then sticks to the play call, or call audible to a different play. Success or failure then creeps into the next play call. Although the OODA Loop was designed for fighter pilots, the concepts within it apply at each level of war, tactical, operational, and strategic.

As discussed previously, the American way of war is to fight jointly, applying power and effects from all domains (air, land, maritime, space, and cyberspace). Landpower and airpower work in concert with sea power. Strategists who think about war, even those in the army or air force must understand the theory behind the navy's application power at sea. This chapter would not be complete without discussing maritime theorists.

4.8 Mahan

Alfred Thayer Mahan was an American naval theorist who taught at the U.S. Navy War College. His theory generally rested upon three principles. First, for a nation to rise to the status of a great power required a strong and powerful navy. In this respect, Mahan drew a direct link between economic and maritime power.[35] Second, protecting friendly commerce and blockading enemy commerce was a path to victory. Third, should the fleet be required for combat, the fleet should concentrate all its combat power and force in a single decisive battle.[36] This final point gave rise to the maxim "Never divide the fleet." Mahan's theories stood the tests of its day, but by in large have not stood the test of time.

The idea that a navy or any military force should concentrate all its forces for a single decisive battle does not hold in modern warfare. At the time of Mahan's writings, ships could not see over the horizon as airplanes, radar, and space-based satellites had yet to be invented. In modern day conflict the ability to operate a dispersed force that can concentrate effects has more relevance than Mahan's theories. The Japanese attempt to knock America out of the Pacific War with the attacks on Pearl Harbor was an attempt at a single decisive battle but failed to consider the strategic impact of an angry American population.

There are few football analogies or that can help to describe Mahan's theories, we can only shoot down the flawed thinking on the idea of decisive battle. Only to say that teams don't earn championships in one single victory. Teams at all levels must play through a regular season, win enough to earn a spot in the playoffs, and then play a series of single-elimination games on the way to the championship game. There is no bye strait to the championship, rather it is a sustained campaign throughout the season.

4.9 J.C. Wylie

Admiral J.C. Wylie, in his book *Military Strategy: A General Theory of Power Control* formulated a theory that the purpose of strategy was to establish some form of control over an adversary.[37] This element of control can be over either the will or the ability to act. To gain control, Wylie postulated that one could formulate either a cumulative or sequential strategy. The cumulative strategy becomes one of attrition and the sequential strategy is one of maneuver.

In his writings, Wylie provides various historical examples of the application of cumulative and sequential strategies. The German attempts to control the Atlantic with U-boats in both World Wars were failed attempts to win by attrition.[38] Wylie uses MacArthur's campaign through the Pacific Islands as his example of a sequential strategy. By moving (and hopping) from one island to another, his forces were able to position themselves closer and closer to the Japanese homeland.

Time of possession is a measurement or way to establish control over an opposing football team. The ability to put points on the board is paramount, but no more than keeping your opponent from putting points on the board. Although not the definitive indicator of victory in a football game, keeping an opponent's offense off the field certainly increases the odds of success. Indeed, the strategy of control is not just the ability to control an aspect of war, but to deny control to the enemy. Although a defense can score on an interception or fumble return, or the occasional safety, the overwhelming majority of points scored in a football game occur on offense. This ball control theory came to light in Super Bowl XXV when the underdog New York Giants defeated the Buffalo Bills. Although the game ended on a missed Scott Norwood field goal, the ending came because of the Giants holding possession of the ball for 40 minutes and 33 seconds, a Super Bowl record.[39]

Of course, there are other methods to attain control, and they don't always necessitate direct combat against the enemy. For example, in recent years China conceptualized the "three warfare strategy" as an approach to attain control. These methods include psychological warfare, media or public opinion warfare, and legal warfare. Using these three techniques, China seeks to change the rules of the game to prevent the United States from employing its strategic advantage of global power projection.

A deep understanding of the rules of the game often leads to a decisive advantage in football. Coaches such as Bill Belichick and his protégé / former player Mike Vrabel often use the rules to their advantage in various schematic formations, or in creative ways to run down the clock towards the end of the game. Moreover, when teams find themselves at a disadvantage, they often press in the off-season for rule changes in the league. Indeed, following Tom Brady's knee injury in 2008, the NFL changed the rules to protect quarterbacks from low hits.

4.10 Conclusion

There is no shortage of military theory to explain wars of the past, present, or future. Theory is useful to think about war, just as it is useful to think about football. The important aspect is not to turn theory into a checklist or maxims in bullet points as a prescription to victory. Military leaders and football coaches should be versed in multiple theories and apply each of them to relevant situations. This means reading and studying more than doctrine and playbooks, but rather an intense study of the history of each respective profession. Which theorist applies in each unique situation can vary based on terrain, the time available, the forces available, and of course the enemy who always maintains a vote in how a battle or campaign will play out. The larger the base of knowledge for the war or football practitioner, the greater the chance of outthinking, or recognizing patterns faster than the adversary. At tactical, operational, and strategic levels flexibility is paramount to solving the problems facing military or football organizations.

Notes

1. Sun Tzu, *The Art of War*. (Las Vegas, NV: Filiquarian Publishing, 2007).
2. Kathryn Watson. "Fight against ISIS has shifted to "annihilation tactics," Mattis says." CBS News. 28 May.
   2018. Accessed 27 November 2018. Retrieved from: https://www.cbsnews.com/news/fight-against-isis-has-shifted-to-annihilation-tactics-mattis-says/
3. Carl von Clausewitz, *On War*, eds. Michael Howard and Peter Paret (Princeton, NJ: Princeton University Press,
   1989), 3.
4. Carl von Clausewitz, 595–596.
5. *Joint Publication 5–0: Joint Planning.* 1 December 2020. Washington, DC: The Joint Staff. Pg GL-6.
6. In U.S. military doctrine, a Center of Gravity contains subordinate elements of Critical Capabilities, Critical Requirements, and Critical Vulnerabilities. After identifying an enemy and friendly COG, planners look for these elements to attack the enemy COG and protect their own COG.
7. Ron Jaworski, *The Games that Changed the Game*. (New York, ESPN books. 2010).
8. Carl von Clausewitz, 6.
9. Cathal J. Nolan. *The Allure of Battle; A History of How Wars Have Been Won and Lost.* (Oxford, UK: Oxford University Press, 2017), 93.
10. Michael McKnight. "Second Nature: Why NFL Coaches are so Prone to Clock Management Blunders." *Sports Illustrated.* Issue 29 August – 5 September 2016.
11. See the drafting of Ryan Leaf or Johnny Manziel.
12. Clausewitz 4.
13. Clausewitz, 114.
14. Mark Fainaru-Wada and Steve Fainaru, *League of Denial: The NFL, Concussions, and the Battle for Truth,* (New York, NY: Three Rivers Press, 2013).
15. Sun Tzu, *The Art of War.* Trans. Samuel B. Griffith. (Oxford, UK: Oxford University, 1971), 84.
16. Sun Tzu, 41.
17. Ibid. pg 93.
18. Ibid, 63.
19. Gordon R. Sullivan and Michael V. Harper. *Hope is Not a Method.* (New York, NY: Broadway Books, 1997).
20. Trent Hone. *Learning War: The Evolution of Fighting Doctrine in the U.S. Navy; 1898–1945.* (Annapolis, MD: Naval Institute Press, 2018). Kindle.
21. Robert B. Strassler, ed., *The Landmark Thucydides: A Comprehensive Guide to the Peloponnesian War* (Free Press, 1996): 350–357.
22. Strassler, 110–128.
23. Strassler, 365–380.
24. Bill Walsh, *The Score Takes Care of Itself: My Philosophy of Leadership.* (New York, NY: Penguin Group, 2009). Kindle.

25. Antoine-Henri Jomini, *The Art of War*. (Project Gutenberg Ebook edition, 28 September 2004), 70.
26. Amy Trask, *You Negotiate Like a Girl: Reflections on a Career in the National Football League*. (Chicago, IL: Triumph Books, 2016).
27. Giulio Douhet, *Command of the Air*. (Pickle Partner Publishing, 1942).
28. Other airpower theorists in the early days of airpower include Hough Trenchard and Billy Mitchell.
29. Johnny R Jones,. "Fundamental Truths of Air Power." In *William 'Billy" Michell's Air Power*. (Maxwell AFB, AL: Airpower Research Institute, September 1997), 14.
30. Nick Cullather, "Bomb the Back to the Stone Age: An Etymology." History News Network. 2006. Retrieved from: https://historynewsnetwork.org/article/30347
31. Robert Coram, *Boyd: The Fighter Pilot Who Changed the Art of War*. (New York, NY: Hachette Book Group, 2002).
32. Doug Farrar. *The Genius of Desperation. The Schematic Innovations That Made the Modern NFL*. (Chicago, IL: Triumph Books, 2018).
33. Michael Lombardi. *Gridiron Genius: A Master Class in Winning Championships and Building Dynasties in the NFL*. (New York, NY: Radom House LLC, 2018).
34. By sending a player in motion, the quarterback can determine if the defense is playing man or zone coverage. Should a defensive player move with the motion, they are in man coverage. Should the defense not react, they are in a zone type coverage.
35. Crowl, Philip A. "Alfred Thayer Mahan: The Naval Historian." In *Makers of Modern Strategy: from Machiavelli to the Nuclear Age*, edited by Peter Paret. (Princeton, NJ: Princeton University Press, 1986), 455–456.
36. Ibid.
37. J.C. Wylie. *Military Strategy: A General Theory of Power Control*. (Annapolis, Md: Naval Institute Press, 1989), 162.
38. Wylie, 25.
39. Pro Football Reference Website. Retrieved from: https://www.pro-football-reference.com/box scores/199101270buf.htm

Chapter 5
Football and U.S. Military Strategy

> Games aren't always won by the more talented team; they are won by the team that recognizes favorable matchups and exploits them. –Ron Jaworski

> "Strategy is the art of making use of time and space. I am less concerned about the latter than the former. Space we can recover, lost time never. – Napoleon

There is no single agreed upon definition of strategy. Contemporary strategist Colin Gray defined strategy as "the direction and use of force for the purposes of policy as decided by politics…the bridge between political purpose and military power."[1] Retired Army Chief of Staff Gordon Sullivan in his acclaimed book *Hope is Not a Method* wrote that "Strategy is an intellectual construct linking where you are today with where you want to be tomorrow in a substantive, concrete manner."[2] Further, John Lewis Gaddis stated that strategy is "the alignment of potentially unlimited aspirations with necessarily limited capabilities."[3] Other definitions of strategy include "strategy is a theory of success,"[4] and "strategy is organized coercion and it can be defensive or offensive."[5] No matter what definition a military commander, head coach, or other strategic leader chooses, they must articulate the strategy (Fig. 5.1).

The American way of war expresses strategy as a 3-legged stool. The legs of the stool represent ends, ways, and means. As such, we can define strategy as a mathematical equation of Strategy = Ends + Ways + Means. Any imbalance between the three is where a nation or military assumes risk.[6] This idea was the formulation of Art Lykke and is the current model taught at U.S. military war colleges and service schools. Football concepts in building teams operate in a similar construct. The owners and general managers determine the resources through personnel acquisition; coaches put systems in place (concepts), with the ultimate objective or end state being a championship. If offensive or defensive systems are not aligned with the capabilities of players, this creates an element of risk. For example, a team with an average quarterback and a group of slow wide receivers will likely have a run-heavy offense. Arguably, the greatest professional football dynasties of recent decades, the 1980s San Francisco 49ers and 2001–2019 New England Patriots consistently balanced their exceptional talent with respective offensive and defensive systems.

D. Sukman, *American Football and the American Way of War*,
https://doi.org/10.1007/978-3-031-55345-5_5

Fig. 5.1 There is no single definition of strategy

Similarly, if the ends or objectives of national security are not aligned with the concepts (ways) or resources (means) allotted to the military and other elements of national power, there is a risk to the likelihood of achieving said ends. If a strategy is not funded, it's not a strategy, it's a fantasy. Indeed, as John Lewis Gaddis explains in his book *On Grand Strategy*, "ends exist in the imagination, they can be infinite…Means, though, are stubbornly finite: they're boots on the ground, ships in the sea, and the bodies required to fill them."[7] This maxim holds in professional football just as it does in warfare. Indeed, just as a nation must budget for military capabilities to fight and win wars, so must general managers and team owners spend money to acquire talent to achieve their ends (Fig. 5.2).

Of course, just as weaker nations with small budgets can defeat a larger military, so can small-budget football teams who acquire lesser talent defeat teams that spend up to the limits of the salary cap. How much money a nation or team spends is a weak indicator and a poor predictor of outcomes. America vastly outspent North Vietnam, Iraqi insurgents, and the Taliban in each respective war, yet never attained a decisive strategic victory. Similarly, on paper, the 2007 New England Patriots, a team that ran the table going undefeated in the regular season, were a stronger team than the New York Giants who squeaked into the playoffs with a wildcard birth. On paper, the Super Bowl matchup between these two teams certainly favored the Patriots. But

Fig. 5.2 Lykke's 3-Legged Stool (S)ource Arthur Lykke, "Defining Military Strategy = E + W + M," Military Review, Vol 69, No 5. (1989))

in football like war, there are no certain outcomes. The ways, methods, or concepts are just as paramount as resources and must align accordingly.[8]

5.1 Focus of Strategy

Military strategies and plans tend to focus and orient forces on one of two objectives. The first is terrain, the second is the enemy. In his book *Military Strategy: A Very Short Introduction*, Antulio J. Echevarria uses the U.S. Civil War to explain the differences. General Grant, throughout his Overland campaign and drive to Appomattox focused his army on destroying Lee's Army instead of seeking to capture and occupy Richmond. On the other hand, General Sherman focused his army on the terrain, destroying rebel infrastructure including railroads and other transportation hubs in his March to the Sea.[9] Each of these strategies has its root in strategic theory discussed in the previous chapter with thoughts on destroying the enemy army espoused by Clausewitz and the importance of terrain emphasized by Jomini. Of course, in war there is no prescription or strategy checklist to follow that will guarantee success, and even the most thoughtful strategies have at least a 50% chance of failure.

The conduct of military strategy often comes to a decision between an offensive or defensive focus. Nations and armies must decide if they will carry the war to the enemy or wait until the enemy exposes itself in the open. Historically, the American way of war and the current preferred method is maneuver warfare with an emphasis on the offensive. Moreover, the American way of war tends to favor playing away games and fighting an enemy or adversary on their home turf. Indeed, following the terrorist attacks of 9/11, the United States rapidly deployed forces halfway around the planet to commence offensive operations in Afghanistan, and two years later continued the offensive in a misguided invasion of Iraq.

More than a focus, strategies tend to have a central idea (or big idea) running throughout. With a central idea, each headquarters is better able to understand the strategy and their role within it. Football teams often operate with a central idea in their game planning. The central idea may be to focus on a single player on the opposing team or to control the clock through long and methodical drives. The big idea must address the central problem a nation, a military force, or a football team faces, and thus drive the strategic direction.

5.2 Types of Strategy

Military strategists and planners have various strategic options they can employ in the development of their theory of victory. These strategies include annihilation, dislocation, attrition, exhaustion, coercion and compellence, deterrence, decapitation or targeted killing, and insurgency and terrorism.[10] Each comes with its innate advantages and disadvantages, but more importantly, there is more than one way to win, in both war and football. The key is to understand the capabilities one possesses and to choose the most appropriate strategy that maximizes said capabilities.

Attrition is the continual grinding down of an adversary's strength, both in people and in material capabilities. Attrition occurs over an extended period under the premise that the opposing side will eventually lose its will to fight. The North Vietnamese successfully employed this strategy against the French and later the United States in the 1950s, 60 s, and 70 s. Exhaustion is like attrition and is primarily wearing down the enemy's will to fight.[11] In this sense, insurgents in Iraq and Afghanistan combined attrition with exhaustion to gradually wear down American patience with each war. Interestingly, the strategy of attrition is not simply a numbers game. The North Vietnamese, the Taliban, and Iraqi insurgents all suffered more casualties than the United States in each respective conflict. Attrition was centered on America's will, not on the fielded forces.

Attrition isn't always the strategy of the insurgent. In his book *How the War Was Won*, Phillips Payson O'Brian describes the U.S. air campaigns in Europe and the Pacific. According to O'Brian decisive to U.S. victory was the ability of U.S. airpower to destroy Axis production over a continual strategic and operational level bombing campaign.[12] The United States, according to O'Brien was able to attrit German and Japanese combat capabilities, and mostly before they ever reached the battlefield. The destruction of production facilities and transportation networks enabled the U.S.

and its allies to bring more combat capability to the battlefields. Attrition is more than killing front line troops.

An **attrition and exhaustion** strategy in football occurs when a team uses its physical strength to wear down the other team throughout a game. Continually attacking and hitting key players on the opposing team (without the intent to injure) will physically wear down the other team. The Giants' strategy against the Buffalo Bills in Super Bowl XXV, and the New England Patriots' strategy against the St. Louis Rams in 2001 employed this strategy. Continual physical punishment of the Bills and Rams offensive stars wore them down, leading to the underdog Giants and Patriots winning the Super Bowl.

Attrition doesn't just happen during a single game. Football seasons are months long. Virtually no team ends the season with the same starting roster as when they began. Injuries to key players add up turning a pre-season championship favorite into an also-ran contender. Indeed, it is teams who have depth at each position and can replace some of the original starters who often end up on the winner's stand.

Decapitation is simply the targeting and killing of an enemy's strategic leadership. This strategy has been popular over the past couple of decades and included the attempted killing of Saddam Hussein on the opening night of the Iraq War in 2003. The United States again used the decapitation strategy in January 2020 when it killed General Qasem Soleimani, Iran's most powerful military commander. The Israelis have a long history of targeting enemy leadership such as the 2004 killing of Sheik Ahmed Yassin, the founder and spiritual leader of Hamas.[13] While David's killing of Goliath might represent an early form of success in winning a war with a precise and targeted killing, modern day war is often more complicated. Comparisons to the popular carnival game "Whack-A-Mole" have come to define U.S. military operations that kill individual enemy or terrorist leaders, who are often replaced by someone just as if not more dangerous.

There is ample evidence of the fallacy or incompleteness of the strategy of decapitation in recent years. As cited in the previous paragraph, the killings of various leaders of Hamas failed to prevent Hamas from carrying out the 7 October 2023 attacks against Israel, igniting the powder keg in the Middle East. Iran still sponsors international terrorism despite removing Soleimani from the chessboard. Indeed, the capabilities and will of a nation or organization tend to be greater than a single individual (Fig. 5.3).

The intellectual foundation behind decapitation is perhaps expressed best in John Warden's 5 Rings. This air power theory binned the capabilities of a state into five categories of rings. On the outer ring are fielded forces of the military, followed by the population, infrastructure, and system essentials leading to the enemy's strategic leadership in the bullseye. Warden postulated that by attacking the center of the ring, U.S. airpower could end the war right at the start.

There are football teams who rely on key individuals for their success. Often this is the quarterback, and at times a running back, wide receiver, or other skilled position. Football teams often focus their largest hits on these players in the hope of knocking them out of the game.[14] Without key players on the field, a team's game plan which

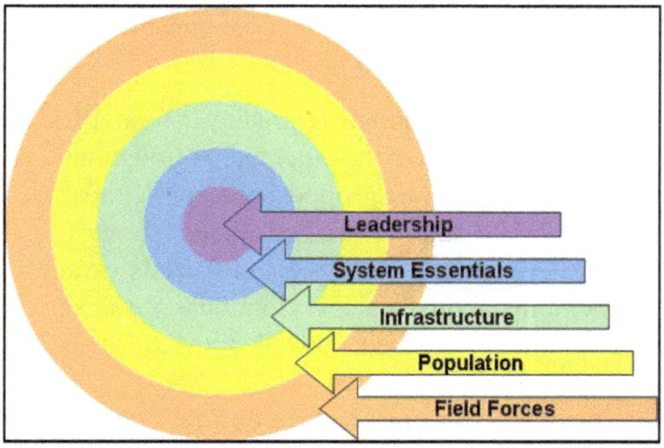

Fig. 5.3 Warden's five rings (*Source* Image available in the public domain on wikimedia commons)

may rely on the unique skills of said player is thrown into disarray. Understanding that k)nocking an opposing player out of a game offers an advantage led to controversies such as the infamous "Bountygate," which resulted in the suspension of New Orleans Saints head coach Sean Payton. In this event, players were offered monetary rewards for injuring opposing players, which is, to say the least, dirty, unsportsmanlike, and against league rules.

While decapitation aims to kill the strategic leaders of an enemy, a strategy of coercion seeks to convince adversary leadership to take or not to take certain actions.

Deterrence, while certainly a legitimate strategy is hard to measure. Often, determining why an adversary chose to take or not take a course of action. Underlying the theory and strategy of deterrence is that it is a strategy that relies on rational calculation. Further, deterrence requires an understanding of what an adversary holds dear, and the ability to place what they hold dear at risk.

In 2022, the U.S. Department of Defense published the National Defense Strategy with a central theme of or central idea of Integrated Deterrence. This, according to the Strategy means "using every tool at the Department's disposal, in close collaboration with counterparts across the U.S. Government and with Allies and partners, to ensure that potential foes understand the folly of aggression."[15] The strategy then clearly lays out how deterrence must come with credible actions and the perception of adversaries that the U.S. along with Allies and Partners have a stake or interest in acting to deny or impose costs on adversary aggression.

This may hold well with established governments, however, states led by dictators and non-state actors may not act in what we consider a rational manner (although completely rational in their minds). Indeed, deterrence against democratic states requires the communication of will and intent towards the population who can influence the decisions of elected officials, while deterrence aimed at dictatorships often focuses messaging on the strategic leaders of the state. Indeed, when thinking through deterrence options, often it is the perception of the adversary that matters, and less so one's determination of the odds of success.[16]

Deterrence comes in two forms: punishment, and denial. Punishment is simply when one side fears decisive consequences should they choose to act in a certain manner. For example, one nation may choose not to invade another based on the fear that the invasion will destroy their armed forces, economy, or another vital interest in the state. Deterrence by denial simply means that one side will not act because they understand that the action will not achieve their desired objectives.

Thomas Schelling, the author of the seminal book Arms and Influence described how violence is most successful when it's not used.[17] For deterrence to succeed a nation must demonstrate both the *capability* and *will* to use specific weapons and other capabilities. For example, nuclear deterrence in the Cold War relied on nations who could demonstrate the capability to launch nuclear weapons and the willingness to do so. These two factors must be communicated to the adversary, as secret weapons that an enemy is unaware of do not achieve deterrence. In modern warfare that occurs in space and cyberspace, deterrence in these two domains becomes even harder as once the capability is known by an adversary, the adversary takes steps to render the space or cyber capability ineffective.

In the movie *Dr. Strangelove or: How I Learned to Stop Worrying and Love the Bomb*, the main character Dr. Strangelove lectures the Soviet Ambassador for not revealing the existence of a Doomsday Machine with the quote "Of course, the whole point of a Doomsday Machine is lost, if you *keep* it a *secret*! Why didn't you tell the world, EH?" Even in satire, Hollywood writers understood that effective deterrence includes communication of capabilities to the enemy.

More than a demonstration of capability and will, proper deterrence rests on effective communication of the two elements. An adversary that does not understand or cannot comprehend the other side will not be deterred.[18] Indeed, effective communication of deterrence messages is paramount when trying to prevent an adversary from choosing a course of action.

Deterrence on the gridiron comes in the form of players with unique skills. Generational talent at certain positions will force an opponent to either focus their talent to neutralize the opposing player or avoid them on the field. An offense may throw to one side of the field for example when the defense includes a cornerback of extraordinary skill. This was the case with the late Seattle Seahawks and Richard Sherman and the Jets and Patriots with Darrelle Revis in the late 2010s. For several years quarterbacks simply saw too much risk throwing the ball in their direction, as it would often result in an incompletion or an interception. Defensive linemen such as J.J. and T.J. Watt, Aaron Donald, and Nick Bosa consistently draw double and triple teams from the offensive line or have opposing teams run the ball run to the opposite side of the field. Opposing teams must develop a game plan and theory of victory that account for players of this caliber who limit the choices of an opposing offense.

Compellence is based on the ability to inflict pain upon an adversary. States that employ this theory must consider the enemy's willingness to endure pain. The ability to inflict pain and suffering upon an enemy's military or an enemy's society is a key capability for a military force.[19] Much like deterrence, coercion requires an understanding of an adversary's interests. Further, while deterrence can have an infinite timeline compellence and coercion have set time limits for an adversary to act.

A football team can compel an opponent into choosing specific plays by shutting down specific formations or plays early in a game. For example, a team may have to resort to passing the ball should a strong defensive front continually stop running plays of the offense. In a similar manner, an offense can force a defensive opponent into various coverage schemes by exploiting defensive formations. A quarterback picking apart a defense aligned in zone coverage can force the defense into a man-on-man package, while a quarterback with a quick release can limit the number of times a defense attempts to execute a blitz.

Insurgency: In every conflict, be it on the battlefield or the gridiron, there is a favorite and there is an underdog. Whenever one conventional force completely overpowers another, the weaker power may turn to insurgency as a method to defeating the stronger power as facing off on the conventional battlefield would be nothing short of suicide. Indeed, insurgents tend to avoid large battles and adopt a long-term defensive strategy. There may be short offensive actions such as raids and ambushes, however, the insurgents will retrograde from the engagement once overwhelming enemy conventional forces arrive where the action is.

There is an old trope that any team can beat any other on any given Sunday. Weaker teams on paper always have a chance to defeat stronger teams given the right game plan. While in many instances at lower levels of football, the size of the community or school influences that available talent level, at higher levels there are always competitive advantages a seemingly weaker team can find to defeat a stronger team. Defeating the larger and more powerful team might mean employing more misdirection or trick plays or taking more risks throughout the game such as not punting on 4th down or going for two points after each touchdown. Creativity is often the key to victory.

5.3 Turning a Strategy into a Plan

War is a combination of art and science. Commanders at each level of war combine the two in the formulation and execution of their war plans. Military planning, which combines elements of the art and science manifests in U.S. military doctrine in the various planning processes. These include the Joint Planning Process (JPP), the Military Decision-Making Process (MDMP), and the Joint Planning Process Air (JOPPA). Moreover, commonly proceeding each of these planning processes is operational design, which helps a commander and staff to understand the tactical, operational, or strategic problem in front of them.

Intelligence preparation of the battlefield (IPB), or Intelligence Preparation of the Operational Environment (IPOE) is paramount to military planning. Indeed, it is unwise to develop a strategy or plan without an understanding of the adversary one will face. Leaders at all echelons of war seek to understand the environment and their enemy. U.S. military doctrine breaks IPOE into four categories, collection, analysis, processing, and dissemination. Dissemination is particularly important, as leaders across the battlefield must have access to the analysis of intelligence. This aspect of planning occurs in football, through scouting. From High School through professional football, teams invest time and resources to watch films of opposing teams to develop their schemes come game day.

Game planning in football relates most to course of action development (COA DEV), a step found in each of the military service's planning processes. This is the aspect of war planning where commanders and staffs execute art and science. Indeed, it is in this phase of the planning process that a staff begins to draw arrows on their maps representing the axis of advance of friendly forces, like the arrow diagrams found inside a playbook. Creativity is essential for the employment of military capabilities, but courses of action must still fall in the realm of the possible. Indeed, each course of action must meet a minimum set of criteria of being feasible, acceptable, distinguishable, adequate, and complete.

Following course of action development, military planners conduct wargames to test their plans, tease out weaknesses within the plan, and improve the plan based on wargaming results. Wargames can range from a simple tabletop exercise (TTX) comprising only commanders and key members of the staff to large scale exercises involving thousands of participants and combat platforms in each of the five domains (air, maritime, land, space, and cyberspace). In a similar aspect, football teams conceive their plan and test it in the week leading up to the game. Testing game plans often runs the gamut from coaches talking through concepts and ideas in front of a whiteboard to full contact scrimmages against a scout team. Well run scout teams can challenge the game design, point out flaws in the plan, and allow coaching staffs to mitigate the identified risk.

Military staffs assess the success and failure of all actions on the battlefield and at each level of war. Assessment in football parallels those of military formations, as players, coaches, and general managers continually assess on field performance. Proper assessment in military organizations involves the entire staff. According to U.S. military doctrine, assessment is meant to "provide perspective, insight, and the opportunity to correct, adapt, and refine planning and execution to make military operations more effective."[20] Assessment in both war and football occurs dynamically, as tactics and game plans can adjust on the fly.

5.4 Timing

An important aspect of the strategy and the operational art is the arranging of actions in time and space. Both war and football have a duration, or a specific amount of time from beginning to end. However, in war, the duration is often ill-defined.[21] Indeed, wars such as the Six-Day War between Arab States and Israel and the 100-Years War between the kingdoms of England and France highlight the unpredictability of time. As recent as 2022, Russian leadership expected to win a decisive campaign in approximately three days over Ukraine. This war would extend over multiple years at a cost of hundreds of thousands of Russian casualties. In war, nothing is certain.

Concerning time, commanders must understand when their forces are ready for action and choose the best time to strike at an opposing enemy. Ideally, the best time is when one's forces are ready, and the enemy is completely unprepared. However, it is rare when both conditions line up. At the strategic level, finding more time in a conflict is a strategy that is often pursued by the weaker military, and to an extent part of an insurgent and attrition-based approach. Fabian in ancient Rome and Washington in the American Revolution extended the time of their respective wars to wear down their enemies. In 1812, the Russians extended the time of their war against Napoleon by refusing a decisive battle and letting the French Army wither away in Moscow's cold and barren winter. Indeed, just as the Germans would learn over a century later, the farther an army advances, the longer it can take to get back home.

Like war, in football, time is a constant factor that head coaches and players must understand and manipulate, and when the opportunity arises to control the clock, it's best to do so. Coaches must decide when it is prudent to freeze the game clock with a time out, and when to use every second of the play clock.[22] Indeed, the ability to manage a clock towards the end of a half or towards the end of a game is a critical skill of head coaches at all levels of football. Teams look to score in the waning moments of a half or a game leaving little to zero time for an opponent to mount an ensuing drive.

It is not a stretch to say that some coaches earned a reputation for their clock management skills. Super Bowl winning coach Andy Reid developed a sour reputation on clock management based on his team's performance in Super Bowl XXXIX. In this example, Reid's Philadelphia Eagles possessed the ball down 10 points with 5:40 left in the 4th quarter. The Eagles would score a touchdown, but only after a long drive that left little time remaining to complete a comeback.[23] These skills or decisions include when to call a timeout, when to run the ball or pass the ball in the middle of the field to keep the clock running, or when to call plays toward the sideline, allowing the ball carrier to run out of bounds to stop the clock.

Most military plans, from the tactical through the operational to the strategic break up into various phases. U.S. doctrine defines a phase as a "definitive stage of a campaign or operation during which a large portion of the forces and capabilities are involved in similar or mutually supporting activities for a common purpose."[24] For a long time, U.S. joint doctrine divided war plans into six distinct phases, of which phase 4, or the lack of a phase 4 plan became a moniker for the Iraq War. Updated U.S. joint planning doctrine places less emphasis on a rigid 6-phase approach but the method of phasing remains as a key element of operational design.[25]

An important aspect of timing is tempo. The timing of football breaks down into phases, such as the 1st and 2nd halves, the four quarters, the last two minutes of the first half, and the last two minutes of the game when teams operate their two-minute drill. It is in this drill where the time impacts the tempo. As time runs to a close, teams often play at an up-tempo pace with the use of the no-huddle and hurry-up concepts. The faster tempo is a direct lift of Boyd's OODA Loop discussed in Chapter 3. Teams use time and tempo to their advantage to make decisions faster than their opponent. However, teams who control the ball while ahead on the scoreboard may seek to slow the tempo of the game. They do this by taking their time between plays hoping to run off as much time as possible to prevent the other team from making a comeback. A decisive failure to control the tempo and the clock was the Atlanta Falcons blowing a 25-point third quarter lead to the New England Patriots in Super Bowl LI.

Finally, when the score is even at the end of the fourth quarter, the game enters a sudden death phase of overtime. In football, the timing of phases is directly related to the clock, while good military planners will base the start and end of each phase based on the conditions of the fight. While military commanders and football coaches alike practice the art of the game, there is a science too. In the military, part of the science is understanding and implementing the principles of war through the application of doctrine.

Notes

1. Colin S Gray, *The Strategy Bridge: Theory for Practice*, (New York, NY: Oxford University Press, 2016), 29.
2. Gordon Sullivan. *Hope is Not a Method*. (New York, NY: Broadway Books, 1997). Kindle.
3. John. Lewis Gaddis. *On Grand Strategy*. (New York, NY: Penguin Press, 2018) Kindle.
4. Jeffrey Meiser and Patrick Quirk, "Want Better Strategists? Start With a Better Definition of Strategy." *Real Clear Defense*. 9 July 2020. Retrieved from: https://www.academia.edu/43964729/Want_Better_Strategists_Start_With_a_Better_Definition_of_Strategy.
5. Michael Howard, *The Causes of War and Other Essays*. (Cambridge, MA: Harvard University Press), 1983, 85.
6. Arthur Lykke. 1989. "Defining Military Strategy = E + W + M," *Military Review* 69, no. 5.
7. Gaddis, Kindle.
8. In a major upset, the Giants defeated the Patriots 17 – 14 in Super Bowl XLII.
9. Antulio J. Echevarria II. *Military Strategy: A Very Short Introduction*. (Oxford: Oxford University Press, 2017). Kindle.
10. Echevarria II.
11. Ibid.
12. Phillips Payson O'Brian, *How the War was Won: Air-Sea Power and Allied Victory in World War II*. (Cambridge, UK: Cambridge University Press, 2015).
13. The killing of various leaders of Hamas failed to eliminate the capability or will of Hamas, as evidenced by the attacks of October 7th, 2023.
14. Of course, teams don't try to deliberately injure another player, and there is a fine line between knocking someone out of a game from hard and legal hits against cheap shots with the intent to injure an opponent.
15. Department of Defense, *2022 National Defense Strategy*. pg IV. Retrieved from: https://media.defense.gov/2022/Oct/27/2003103845/-1/-1/1/2022-NATIONAL-DEFENSE-STRATEGY-NPR-MDR.PDF.
16. Michael J. Mazarr, *Understanding Deterrence*. (Santa Monica, CA: RAND, 2018).
17. Thomas C. Schelling, *Arms and Influence*. (New Haven, CT: Yale University Press, 1966), 10.
18. Lawrence Freedman, *Deterrence*. (Malden, MA: Polity Press, 2004), 28.
19. Schelling, 2.
20. *Joint Publication 5–0: Joint Planning*. (Washington, DC: The Joint Staff, 1 December 2020), Pg VI-I.

21. Robert R. Leonhard. *Fighting By Minutes: Time and the Art of War*. (KC Publishing. Kansas City, MO, 2018).
22. In professional football, the offense has 40 seconds to snap the ball following the previous play before they are penalized 5 yards for a delay of game.
23. Andy Reid however has more than made up for this in his long and distinguished career having his team advance to and win multiple Super Bowls in later years with the Kansas City Chiefs.
24. Department of Defense Dictionary of Military and Associated Terms. (16 February 2016), 184.
25. Joint Publication 5–0. Joint Planning.

Chapter 6
Levels of War

> As far as Saddam Hussein being a great military strategist, he is neither a strategist, nor is he schooled in the operational arts, nor is he a tactician, nor is he a general, nor is he a soldier. Other than that, he's a great military man, I want you to know that. – General Norman Schwartzkopf.
>
> The game of football is one of strategy and tactics. – Woody Hayes

Like war, football occurs on three levels. The United States military recognizes three distinct echelons of war. At the lowest rung is the tactical level, followed by the operational level, and culminating with the strategic level on top. American football is a useful analogy to the three recognized echelons of war and helps us to explain why winning at the tactical level—while immediately satisfying—is not the goal of football teams, or of nations at war.

U.S. military doctrine defines the tactical level of war at where battles and engagements are planned and executed to achieve military objectives assigned to tactical units or task forces.[1] Although there is doctrine for each level of war, doctrine is particularly important to tactical military units. To fight as a unified force, squads, platoons, companies, battalions, brigades, and divisions or ships, squadrons, and fleets turn to doctrine to guide their actions in combat. Doctrine provides a common language or lexicon and creates a shared understanding between commanders on what actions to expect of their subordinate, adjacent, and higher headquarters. These actions include what to do when in close contact with the enemy and how the unit will maneuver forces on the battlefield.

In the military, the overwhelming majority of service members (both enlisted and officers) will spend most if not all of their careers at the tactical level. This can be frustrating, as service members begin to observe that tactical success doesn't always translate to strategic victory, as the best tacticians can transform into average or pedestrian strategists. War is more than the cumulative effects of tactical actions, and tactical victory after tactical victory does not always lead to national level success. The Vietnam War is always a useful metaphor for this phenomenon, as the U.S. military never lost a tactical fight. Tactics without strategy is the noise before defeat.

The tactical level of football is the game between the sidelines. The actions of the players on the field, executing called plays, constitute the tactical aspect in the

D. Sukman, *American Football and the American Way of War*,
https://doi.org/10.1007/978-3-031-55345-5_6

same manner that soldiers on the ground execute their battle drills. The game is divided into three tactical units, the offense, defense, and special teams. Further, at the tactical level in both war and football, training and preparation before execution is paramount to success. Winning at the tactical level is winning a single game. It is here that a critical difference between football and war becomes noticeable. The cumulative effects at the tactical level in football tend to generate operational and strategic effects. Single game victories in a season add up to a playoff birth, where single game victories can lead to a season championship. Moreover, tactical success tends to generate more fan interest leading to more revenue for a franchise at the strategic level.

The operational level of war is where campaigns and major operations are planned, conducted, and sustained to achieve strategic objectives within theaters or other operational areas.[2] The operational level of war came to fruition in the more modern era of warfare as armies and navies could not win a war with one decisive battle. Military leaders had to link battles and other actions together in a comprehensive military campaign. Much of the thought behind the operational level of war emerged following the Napoleonic Wars where the outcome of a war often occurred on a single day and on a single terrain feature.[3] For example, neither the United States nor the Confederates in the U.S. Civil War could defeat the other in a single victory. Though General Winfield Scott's conception of the Anaconda Plan displayed operational level thinking, it was not until Grant took charge and linked battles in the American West, the South, and the North that the U.S. was able to attain decisive military victory over the totality of the Confederate military.

If military commanders at the operational level focus on the application of time, space, and forces available, then the operational level of football lies with the coaching staff. Developing schemes to orchestrate the efforts of their tactical formations (offense, defense, and special teams) is the responsibility of a host of coaches, including offensive and defensive coordinators, strength coaches, and position coaches. Further, the coaching staff is responsible for organizing the team on the battlefield and finding opportunities to exploit the opposing team's weaknesses. This is like operational art. The ability to arrange forces on the battlefield and to conduct proper Intelligence Preparation of the Operational Environment (IPOE) such as scouting an opposing team, are essential elements in the lead-up to the weekly game. Winning at the operational level in football means linking victories throughout the season to win a division, conference, or league championship.

In both football and war, it is the operational level where the intellectual begins to outweigh the physical aspect of combat. Leaders at the operational level cannot always lead from the front lines as they must make operational level decisions that can inform the outcome of a campaign. Indeed, when, and where to apply combat power and other battlefield functions such as sustainment and fires is of paramount consideration as not all combat power at the operational level can be applied to a single decisive point.[4]

Like the tactical level of war, operational level victories and operational level brilliance doesn't necessarily translate to strategic success. Indeed, the German military at the onset of the Second World War excelled in operational level warfighting.

The Germans famously combined infantry, armor, artillery, and close air support to rapidly overwhelm the Polish and French armies. However, Germany soon found itself fighting the war on multiple fronts with few allies or partners to share in the warfighting burden. If tactics is the noise before victory or defeat, the operational level is the instrument that can amplify successful or unsuccessful tactics.

In the military, there has been a general observation that successful leaders at the tactical level don't always succeed at the operational or strategic levels. Tactical genius doesn't equal strategic acumen. Indeed, as most military leaders spend most of their careers at the tactical level, when placed in operational or strategic leadership positions, they sometimes revert to tactical techniques and methods that they are comfortable with. The best captains don't necessarily make the best colonels, and the best colonels don't automatically become competent generals. Similarly, hall of fame caliber players don't always become successful head coaches, and successful head coaches don't always work out in a team's front office as a general manager.

Strategy, according to David Zabecki is where the focus becomes winning wars at the national level.[5] U.S. military doctrine defines the strategic level of war as the level of war at which a nation, often as a member of a group of nations, determines national or multinational (alliance or coalition) strategic security objectives and guidance, then develops and uses national resources to achieve those objectives.[6] Owners and general managers constitute the strategic level of football. Owners possess the institutional responsibility to man and equip a team. This is the owner's equivalent of the military services, who hold *Title 10* (of the U.S. Code) authority to man, equip, and train the force for employment by Combatant Commanders.

General managers, at the strategic level, manage the resources or budget provided by the owner to assemble the best force they can. Moreover, owners and general managers are responsible for developing the acquisition requirements for their respective teams. Owners must hire the right coaching and supporting (e.g., medical) staff, purchase the proper training equipment, and furnish the players with incentives to perform to their highest capabilities on the field of play. Most importantly, general managers must decide which players will consume large portions of the salary cap, be it the quarterback, wide receivers, left tackles, or defensive backs.

The strategic level includes the efforts and activities in building a military. Activities such as marketing and recruiting new personnel into the force, designing and developing new combat systems and platforms, developing new and innovative doctrine all fall under the auspices of building a military. In these efforts, civilian and uniformed leaders must continually choose what timeframes they wish to focus their time and resources on. A continual source of friction is the decision to focus on developing capabilities and spending capital on current conflicts versus spending capital for future force capabilities. This friction came to the forefront at the height of the wars in Iraq and Afghanistan in the 2006–2010 timeframe.

Secretary of Defense Bob Gates came into office as the Secretary of Defense in 2006, following the resignation of Donald Rumsfeld. Gates brought with him the understanding that America's priorities should be the current conflicts. This was different from Rumsfeld's worldview that capabilities development along with funding and investment should focus on the long term and future threats to the United

States and its allies and partners. This latter view centered on high technology to fight peer adversaries such as Russia and China. Ultimately, Secretary Gates changed the culture in the services, forcing acquisition and capabilities development to focus on the current fights in Iraq and Afghanistan such as the Mine-Resistant Ambush Protected (MRAP) vehicles. Strategy is very much a balance or calibration between the needs of today against the needs of tomorrow. Too much focus on counterinsurgency is often the driver of risk when major combat operations against a near-peer adversary lie on the horizon.

Owners, general managers, and head coaches consistently make choices in the design and employment of their respective football teams. Choices range from what players to draft, what round to draft them in, what free agents to sign, and what types of offensive and defensive systems and schemes to employ on the field. Further, informing these choices is how competitive a team is in the short term. A team may sign expensive free agents to a short-term contract in the hopes of bringing home a championship in the near term or rely on draft picks at a lower salary with the understanding that a championship in the near term is out of reach. All these choices are interrelated, as the best teams don't necessarily acquire the best players, but rather obtain the best players that fit into their systems.

More than simply the acquisition of technology, strategic leaders must balance the needs of today's force against the needs of the future vis-à-vis force structure, end strength (the number of service members), and readiness. Known as the iron triangle of painful tradeoffs, a simple rule of thumb is that strategic leaders get to focus on two of these categories at the expense of the third.[7] While there may be no absolute right decision on this balance, there are certainly wrong ones. While coaches and service members at the tactical level think more about the present and the problems they face in front of them, leaders at the operational and strategic levels must look and think further into the future. Whether it is signing a player to a long-term contract or purchasing a new weapons system from Lockheed-Martin, operational and strategic leaders must live and think in the future. It is the future where strategic leaders must live, where they have the greatest impact, and in the future where strategic leaders assume the most risk.

Complicating matters is that the acquisition of combat platforms or the signing of players can commit a military or a team to a system for a long time. Indeed, in the military, combat platforms such as the B-52 bomber, M1 Abrams tank, and ballistic missile submarines stay in the service for decades, often outlasting the service and lifetime of those who initially fielded the system. The B-52 came into service in the early 1950s and is still a key part of the U.S. Air Force's force structure 70 years later. In the case of navy ships, the pace of ship building drives the life cycles of ships to extend to 30 or 40 years.

Matching the right talent to a team's system is certainly a theory of success but becomes complicated when faced with fiscal restraints. For example, general managers must balance investment between offense, defense, and special teams. Indeed, at the professional level, a team must decide on what positions to invest in the best, most experienced, and most expensive talent, and what positions to assign to young and inexpensive players. Deciding to go all-in on a championship run in the current year can fill up salary cap space in future years on players with diminishing levels of talent. At lower levels, coaches must often decide between playing more experienced and older players who might offer a better chance to win today against playing younger and less experienced players in the hope that that investment will pay off in the future. In either choice, the outcome is never certain.

War does not occur in a vacuum and is not the sole domain of the military. While not in U.S. joint doctrine, there does exist a grand strategic level of war. The military historian and strategist B.H.L. Hart conceived of this level as where the coordination of all the resources of a nation occurs.[8] These resources involve all the elements of national power including diplomacy, economic power, information, law enforcement, intelligence, and the military. In the United States grand strategy is the domain of elected leaders who oversee, manage, and synchronize these efforts. The President as the head of the executive branch of government is perhaps the most important individual at this level presiding over the National Security Council. However, as Congress holds the power of the purse and controls what capabilities across government receive funding, elected representatives and senators also have a role. It is Congress that decides the budget for the services, sets the end strength of the services,[9] and votes on authorizations for the use of force, or to declare war.[10] It takes a nation to fight a war.

Just as there is a grand strategic level of war, so is there a grand strategic level of football. This level is the concern of the commissioner of the league who represents all owners. Indeed, the commissioner of the NFL is elected by the ownership and represents the collective ownership of all the teams. As a leader at the grand strategic level, the owner must exercise strategic foresight and anticipate the ever-changing conditions and values of American society to make changes to the game. In recent years this included changing the rules to create more offense and attempts to limit injuries, particularly concussions. Moreover, the commissioner must regulate the off-field behavior of the league's athletes lest they come to a head as in the Ray Rice scenario.[11] Just as all agencies in government don't necessarily agree with each other on how, where, when, and why to employ the various instruments of national power, the commissioner doesn't always have consensus with all the stakeholders impacted by his decision making.

Strategy, whether in war or football is about choices. These choices often come down to money, and where to invest capital. Just as nations are constrained by fiscal budgets in how much they can spend on defense programs, professional football teams are constrained by the salary cap. Although the U.S. Department of Defense operates an annual budget of over $850 billion per year, the decisions on how to spend the money are the result of compromises among multiple strategic leaders.[12] Decisions must be made on what weapons systems, combat platforms, and personnel policies require investment, and what capabilities the Department should divest from. Choices in strategy are not unique to national defense as they are readily apparent in football.

The various long-term choices lead us to the development of the institutional level of war. Tacticians and strategists alike tend to find more glory and fame by drawing arrows on a map, but the most influential officers work behind the scenes to build and maintain the military. Indeed, the fundamental flaw of limiting thinking to the tactical, operational, and strategic levels of war is that it does not consider the many hours of training it takes to create a ready force, the industrial might it take to develop innovative weapons systems, or the years and decades it takes to create a cohort of strategic leaders. Although previously discussed as part of the strategic level within U.S. military doctrine, it is the institutional level of war where we find the creation of warfighting concepts and doctrine, the recruitment and professional development, education, and training of service members from the newest private in basic training to a general officer attending capstone education.

A second level of war not a part of American service or joint doctrine to consider is the technical level of war. It is here where individual actions occur, as does the performance of individual weapons systems.[13] The maintenance of M-4 rifles for example, or the performance of a single tracked or wheeled vehicle. While seemingly irrelevant in the larger scheme, the technical level is where the Clausewitzian friction often begins. A breakdown of a single vehicle in a military convoy can back up traffic on a supply route causing operational delays of larger formations, and of course then serve as a large juicy static target for the adversary.

If there is a technical level of football it occurs at the individual level and with individual items of equipment. Single players learning the proper way to run a pass pattern, quarterbacks developing their footwork, and defensive players perfecting their tackling techniques serve as examples. For equipment, each team often has an equipment manager responsible for maintaining the team's gear. As the football world learned from 2014 to 2016, there are even people responsible for the proper inflation of footballs before a game.[14] Much like the technical level of war, in football friction often begins at this level.

Finally, it is useful to think about policy and the relationship between policy and strategy. Indeed, before the conception of the operational level of war, Clausewitz saw policy as the third level of war alongside tactics and strategy.[15] In the American model, policy is formulated by elected and appointed civilian officials. Moreover, it is in the formulation of policy that we find the conception and in turn communication of a nation's war aims. It is then the job of leaders at the strategic level to develop a strategy to achieve policy goals. A critical aspect of understanding policy is that often the relationship between policy and strategy is an ongoing and dynamic dialog. Success or failure in the military realm, even at the tactical level can change a nation's goals and its subsequent formulation of policy.

When thinking about football and the American way of war, it is useful to look beyond a single play or single game. Professional football teams conduct a 17-week season campaign every year, linking their weekly battles with hopes of culminating in the Super Bowl. The best teams, those dubbed dynasties set the conditions for multiple campaigns through a system that sets the conditions for success year after year. This success occurs when the components of the system at each level work together. Coaches coach and players play while general managers wheel and deal, all to achieve the ends determined by the owner. Dysfunction among them leads to failure. Neither players nor coaches nor owners, can do it alone. Football players and coaches don't win championships, football organizations do.

Soldiers fight and diplomats talk while elected leaders make policy. Armies don't win or lose wars, nations do.

Notes

1. Joint Publication 3–0. *Joint Campaigns and Operations*. (Washington D.C.: The Joint Staff, 2022). Pg 1–12.
2. Ibid.
3. Georgii Samoilovich Isserson, *The Evolution of the Operational Art*. (Fort Leavenworth, KS: Combat Studies Institute Press, 2013).
4. Ibid.
5. David T. Zabecki, *The General's War: Operational Level Command on the Western Front in 1918*. (Bloomington, IA: Indiana University Press, 2018).
6. Joint Publication 5–0. *Joint Planning*.
7. Mara E. Karlin, *The Inheritance: America's Military After Two Decades of War*. (Washington, D.C.: The Brookings Institution, 2022), 111.
8. B.H. Liddell Hart. *Strategy*. Second Revised Edition. (New York, NY: First Meridian Printing, 1991), 322.
9. End strength is simply the number of men and women in uniform. Congress sets the end strength of each service in the National Defense Authorization Act (NDAA).
10. The last time the United States declared war was in 1942, during WWII against Hungary, Romania, and Bulgaria who were allied with the Axis power. Not declaring war has not however, prevented war.
11. Ray Rice is a former running back for the Baltimore Ravens. After accusations of domestic violence, he was suspended for two regular season games. When the violent video of him punching and knocking out his then fiancé and later wife in an elevator was released, the ensuing embarrassment of the light punishment was devastating to the NFL.
12. As of Fiscal year 2023, the U.S. DoD Budget came in at $857.9 Billion. Source: https://www.armed-services.senate.gov/imo/media/doc/fy23_ndaa_agreement_summary.pdf.
13. Edward N. Luttwak, *Strategy: The Logic of War and Peace*. Harvard University Press, 2002).
14. This is a call out to the famous Deflategate involving Tom Brady and the New England Patriots. This saga played out over two years, multiple lawsuits, investigations, and blog posts by world renowned physicists on the intricacies of air pressure and cold weather. Tom Brady would be suspended for 4 games in the 2016 season, only to come back to win his record-breaking 5th Super Bowl title.
15. Michael Matheny. "The Fourth Level of Warfare." *Joint Forces Quarterly* 80, 1st Quarter 2016. Pg. 62–66.

Chapter 7
Operations and the Principles of War

A leader has the right to be beaten, but never the right to be surprised. – Napoleon Bonaparte.

Surprise is often simply a default device, something in lieu of strong thinking. –Bill Walsh.

7.1 American Football and the U.S. Doctrines of War

The operational level of war lies just below the strategic level and sits on top of the tactical level. The operational level of war as previously discussed in the book is where leaders link successive engagements and battles into a higher concept. While poor tactics can create the conditions for a slow and painful strategic defeat, successful operations can accelerate the conditions of strategic success.

In U.S. military parlance, the operational level of war begins in the realm of joint operations, which is where two or more services (Army, Navy, Air Force, etc.…) work together. These operations occur in the five domains of land, air, maritime, space, and cyberspace. It is a joint force commander that synchronizes the actions and effects of each of the forces in each of those domains in space and time and elevates a simple tactical action to the operational level. Linking engagements and battles into a broader operational campaign is not a new phenomenon in war. Indeed, the evolution of the operational level of war and the operational art date back over two hundred years.

Two centuries ago, Napoleon suffered his final defeat at the Battle of Waterloo. Up until that moment, Napoleon revolutionized warfare in his operational approach to battle. Clausewitz would follow the Napoleonic wars with his seminal writing *On War*, in which he described the science of warfare. Just as influential, Baron Jomini would take the lessons he learned while fighting under Napoleon and translate them into Principles of Warfare. These Principles published in *Precis de l'art de la guerre* (*The Art of War*) included maxims such as focusing an attack on a decisive point and the ability "to quickly maneuver and engage fractions of the enemy's army with the majority of one's own."[1] These principles are as applicable to warfare today as they were two centuries ago. In a contemporary context, I offer the principles of warfare

D. Sukman, *American Football and the American Way of War*,
https://doi.org/10.1007/978-3-031-55345-5_7

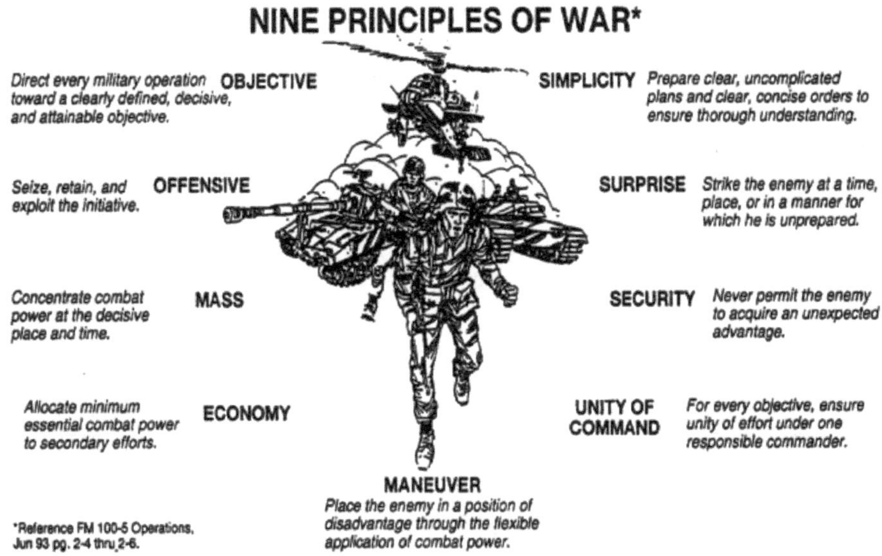

Fig. 7.1 The nine principles of war (*Source* U.S. Army Field Manual 100–5, Operations, June 1993)

as a method to design a higher echelon of strategy vis-a-vis the game of football (Fig. 7.1).

U.S. military, like the American football teams, play by rules; and both are doctrinal. Today, joint U.S. military doctrine recognizes nine principles of war, which include the following[2]:

Objective: To direct every military operation toward a clearly defined, decisive, and attainable objective. Examples of a clearly defined military objective include defeating an enemy force or seizing a key piece of terrain. When a military force is unclear what its objective is on the battlefield, chaos ensues. Arguably, following the fall of Baghdad, the U.S. military failed to define an objective for forces in Iraq, leading to years of strategic wandering. Certainly, every football team begins the season with a clearly defined objective; to win a championship.

At the strategic level, objectives are defined as war aims. Like objectives, the war aims at the onset of a war often morph as the conflict unfolds. For example, Lincoln's initial war aims were limited to putting down the rebellion of the Southern States in 1861 changed to emancipation and ending of slavery following the battle of Antietam. A more recent example is the U.S. missions in Iraq and Afghanistan where the initial objectives to overthrow the Taliban and Saddam Hussein shifted to one of nation building and counterinsurgency. Indeed, war often has a way of going in various and unexpected directions. At the operational level objectives tend to focus on the enemy force or the terrain. The defeat of a specific enemy unit or the occupation of a city or valley are two examples of operational objectives.

Football teams are often realistic in their expectations each year and understand that a championship or playoff appearance is out of reach. This is when a team will enter a rebuilding phase, where the objective is to compete, but also to release veteran players while signing younger players to reduce salary expenses. The poor showing over the course of a season will result in a high draft choice, allowing the team to generate talent at a lower price to compete in the long term.

However, when that objective falls to the wayside early in the season, or when a team enters a season understanding that a championship is a step too far, head coaches and general managers often attempt to change the objective. Often head coaches will term the objective as "rebuilding." This is a false method of hope; and when a football team redefines its season objective of rebuilding in lieu of victory, organizational chaos ensues. Conversely, some teams attain unexpected success early in the season and look to obtain a specific player to meet an unfilled need in the offense or defense to make a run at and in the playoffs.

Offensive: To seize, retain, and exploit the initiative. Clausewitz wrote that defense is the strongest form of warfare. However, the American way of warfare remains offensively focused with the understanding that you cannot win a war without eventually attacking your enemy. There are other ways in which armies attain victory in war. George Washington used a Fabian approach by avoiding direct battle with the British, but ultimately the culminating battle of the Revolutionary War occurred at Yorktown, where the combined forces of the French and the Americans laid siege to General Cornwallis and his army. The tide of the American Civil War did not change until Grant took command of the Union Army and combined his offensive plans in Virginia with General Sherman's offensive through Georgia and the Carolinas. Avoiding battle and remaining on the defensive can buy time, but not decisive victory.

The National Football League has seen an evolution of conflict between the strongest forms of football. From the time the New York Football Giants responded to the Joe Montana-led San Francisco 49ers with a defensive roster that could shut down the West Coast Offense led by the likes of Lawrence Taylor and Jim Burt.[3] Further evolution of the conflict pitted defensive juggernauts such as the 1985 Chicago Bears, 1990 New York Giants, and 2013 Seattle Seahawks against the offensive leviathans such as the 1980s San Francisco 49ers, 1990s Buffalo Bills, and the two-decade run of the New England Patriots. Although the pendulum swings back and forth between offense and defense, the evolution of the game and its respective rules governing the sport tend to favor the offensive attack. In the early 2020s, offensive teams tend to fair better and make deeper runs into the playoffs. The 2020 and 2022 Super Bowl Champion Kansas City Chiefs and 2021 Super Bowl Champion Los Angeles Rams exemplify this offensive trend.

Mass: To concentrate the effects of combat power at the place and time to achieve decisive results. Ensuring sufficient combat power for all phases of an operation is a hard lesson relearned most recently from the wars in Iraq and Afghanistan. Arguably, leaders such as Army Chief of Staff General Eric Shinseki understood this principle when testifying that an occupation of Iraq would require a couple hundred thousand boots on the ground. When troops or combat platforms cannot be massed at a single point of attack, the concept of massing effects becomes the desired course of action. This can include a combination of actions such as cyber-attacks and precision guided munitions taking down adversary air defense systems.

Overpowering an adversary at the point of attack is a concept that easily translates from the battlefield to the football field. In the tactical realm, the ability to either outnumber an opponent—whether it's defenders on the line of scrimmage combining with blitzing linebackers to attack the quarterback or maneuvering offensive lineman to where the football will be on a screen or bubble pass—is a fundamental concept. From the early employment of the V or flying wedge through Vince Lombardi's power sweep to modern day pass routes, the number of players at the position of attack often provides the decisive advantage on the gridiron.

Economy of Force: To allocate minimum essential combat power to secondary efforts. In military operations, most commanders identify a main effort and supporting efforts. The main effort thus receives priority for intelligence, logistics, and fires support. At the operational level, an economy of force was in display with the invasion of Iraq. In 2003, the bulk of U.S. combat forces executed a rapid assault on Baghdad. Other cities in Iraq such as the Southern Port City of Basra required attention and military operations to secure initial victory; however, as a secondary effort, limited capabilities were applied to secure the city and occupation of the city would be the job of the British.

Football teams often focus their strength at certain positions while assuming risk in others. In the era of the salary cap, professional football organizations must determine how to spend their money on top talent while balancing risk with the acquisition of lesser talent. Doling out top dollar for a quarterback and wide receivers may force an economy of effort in the offensive or defensive line, running backs, or other areas of the game such as special teams. Indeed, understanding where to focus efforts to provide a competitive advantage over an adversary is just as prudent in football as it is on the battlefield.

Maneuver: To place the enemy in a position of disadvantage through the flexible application of combat power. In designing operations or strategy, a commander looks to keep his force in key positions on the battlefield taking advantage of time, terrain, and the forces available to each side. Geography matters, and from the time the Continental Army took Dorchester Heights, overlooking Boston and the British occupying forces, to the Left Hook of the Gulf War that led to the rapid destruction of the Iraqi Army, proper maneuver on the battlefield has always been decisive to victory.

Maintaining a relative advantage in field position is one aspect of maneuver. If a team cannot score on every possession, it should continue to place its opponent in a position of disadvantage by having them start an offensive series deep in its own territory. In this aspect, flipping the field through effective special teams play is paramount. Investing in a top tier punter or players who can race down the field on a kickoff or punt is a wise decision, even if it comes at a cost to other areas of the game. Super Bowl LIII where the Rams and Patriots combined for fourteen punts is a prime example of the importance and value teams place on quality field position.[4]

Unity of Command: To ensure unity of effort under one responsible commander for every objective. The best military plans and operations occur under the unified direction and command of a single commander. When military forces operate under separate commands, the result is often duplicative or conflicting.

Today, the American Way of War is by, though, and with, joint operations. Taking in the lessons of World War II, Vietnam, and the invasion of Grenada, in 1986, Congress passed the *Goldwater-Nichols Act*, which created streamlined the chain of command so Combatant Commanders would report directly to the Secretary of Defense and the President.[5] This ensured that all forces from all the services fighting in a military operation would report to one commander. Moreover, the Goldwater-Nichols Act strengthened the position of the Chairman of the Joint Chiefs as an independent military voice for providing advice to the President.

Today, one of the singular decisive advantages of the U.S. military is the ability to fight as a joint force. Indeed, one of the inherent strengths of how the U.S. military fights is its use of a single Joint Force Commander instead of separate service commanders for a given operation. During times of crisis and conflict, from humanitarian assistance operations to major combat operations, the U.S. Military responds with a force operating under a joint force commander. This commander could be the 4-Star level combatant commander or a under an officer serving as a joint task force (JTF) commander. Regardless of the level, all the forces from each of the services operate under a single commander.

In terms of football, the best-run organizations are those with a continuity from the owner, through the general manager, to the head coach, and down to the players on the field. For example, in the 1980s, the New York Giants coaching staff remained consistent for half a decade, to include head coach Bill Parcells, defensive coordinator Bill Belichick, and offensive coordinator Ron Erhardt.[6] Organizations such as the Pittsburgh Steelers and New England Patriots have maintained a level of long-term success with multiple division, conference, and Super Bowl titles. Other examples of unity of command in professional football include the early 1990s Dallas Cowboys under Jerry Jones and Jimmy Johnson, where the players drafted and acquired fit neatly into the system of the head coach, and the New England Patriots under head coach Bill Belichick, where for two decades the acquisition of talent matched the schematic designs of the head coach.

Further, on the field, there is little doubt who oversees the team. While a head coach may delegate play calling to offensive and defensive coordinators, the head coach maintains the authority to step in and determine the play call. Further, some decisions are reserved specifically for the head coach, such as going for it on 4th down, attempting a 2-point conversion, or challenging a referee's call.[7] On the field, teams designate captains of each phase of the game; offense, defense, and special teams. These players are the leaders in the locker room and on the field, but more in the sense of peer leadership. Finally, while the offense is on the field it is the quarterback who is in charge, who calls the play in the huddle, and who is charged with reading the defensive alignment and adjusting the blocking scheme, the receiving routes, or calling an audible and running a completely different play.

Security: To never permit the enemy to acquire unexpected advantage. Security is paramount to any military force. At the tactical level, this could mean avoiding an ambush. At the strategic level, it might be the capacity to protect forces to avoid the disastrous effects of a suicide bomber, as seen in Beirut in 1983. In modern warfare, commanders continually evaluate the security measures of the force within any given operation, from missile defense over an airfield to the convoy security measures taken on the roads in Afghanistan and Iraq.

We see security translated to the football field in the design of a solid defensive scheme. Further, teams consistently take caution on special teams so as not to be surprised by a fake punt, fake field goal, or unexpected onside kick. Further security lies in the capacity of the offensive line to protect a star quarterback. Michael Lewis, in his book *The Blind Side*, detailed the value of the left tackle position in how teams emphasize the position to ensure the center of gravity of a team stays healthy.[8] One team can quickly gain an advantage over another by knocking a strong quarterback out of the game.

The rise of smartphone technology and the rapid pace of data sharing in modern times means safeguarding or securing information and preventing an adversary from gaining access to classified material, thus providing an unexpected advantage. Operational security, commonly called OPSEC in the military is now an aspect of every military plan and operation. Military forces encrypt messages, develop code names for plans and operations, and classify material to prevent distribution. For example, material such as plans and orders are given the label unclassified, secret, or top secret to limit who can receive and read them. Further, when military forces embark on a mission, there are typically strict requirements and limitations to what a service member can and cannot speak to in the public sphere. In modern times, military commands restrict the use of cell phones and other electronic devices that an adversary can monitor and exploit for data.

In all levels of football, teams go to extraordinary ways to ensure other teams cannot predict what they will do on the field. These measures include quarterbacks and coaches covering their mouths while issuing a play, developing a system of signals from the sideline to the players on the field, and speaking in a coded language that players must re-learn each time they join a new team. More recently, college football coaches have used photo boards with various images from cartoon figures to pop culture icons to call signals onto the field.[9] Moreover, there are specific rules that guide the way teams can scout another, to include strict limits on what can and cannot be videotaped (Fig. 7.2).

Surprise: To strike the enemy at a time or place or in a manner for which it is unprepared. In terms of military operations, two historical examples stand out. First, the 1991 "Left Hook" executed by U.S. ground forces in the liberation of Kuwait. Saddam Hussein expected an amphibious assault. As the air campaign began, his regime's ability to monitor U.S. forces disappeared. American leaders took advantage of this by posturing forces for an offensive at an unexpected (by Hussain) location behind the Republican Guard. More than a creative plan, surprise can come in the form of employing new weapons and technology or employing old technology in

Fig. 7.2 Photo boards in College Football (*Source* Si.com "Funny photo boards used to call signals from the sideline." October 10 2015)

new and creative ways. Generally speaking, Al-Quada's use of airplanes to attack the American Homeland on September 11th caught the United States by surprise.

In terms of football, teams look to achieve surprise through game planning, and the study of an adversary in the week prior may lead to a surprise tactic during a game. More than a single gadget or trick play, achieving surprise on an opponent may include creating a mismatch on the offensive or defensive line, using a ground attack when an opponent is expecting a passing game, or setting up in a new, original, and unexpected formation.

One way to achieve surprise is to build up a tendency for action to condition an opponent to a certain perspective. For example, the use of snap exercises by Russia in the lead-up to the invasion of Ukraine in 2014 diverted attention from Russia's true intentions of annexing Crimea and occupying the Donbas region of Eastern Ukraine.[10] Add in Egypt's exercises before the 1973 Yom Kippur War, and we can see how a nation and its respective intelligence agencies can become desensitized towards what would otherwise be indications and warnings of an imminent invasion. Adding to the desensitizing through exercises is the cover of current events. Russia now has a history of using the Olympics as a distraction for their military build-ups prior to invasions as they did before invading Georgia in 2008, and Ukraine in 2014 and 2022.

We see this aspect of surprise when football teams continually run the ball, then mix in a play-action fake to draw defensive backs closer to the line of scrimmage. Quick counts or fast snaps after using the same over the course of a game can catch a defense off guard. On the other hand, a hard count where the quarterback raises his voice while calling signals may lure the defense offsides resulting in a free five yards. Defenses at times will drop a lineman into pass coverage while blitzing a defensive back to catch the offensive line and quarterback off guard. Teams often get lulled into a routine expecting certain formations and personnel packages to signal specific plays, the best teams capitalize on this feature of human nature. Other elements of surprise are the employment of previously unused formations, such as the Miami Dolphins' use of the wildcat formation in 2008. Football tends to be a copycat sport and new and surprising formations are only new and surprising for so long before other teams employ the tactic.

Simplicity: To prepare clear, uncomplicated plans and concise orders to ensure thorough understanding. Violating the principle of simplicity in warfare creates conditions where commanders on the battlefield may be unclear of a higher commander's intent or scheme of maneuver. War is the most complex of human activities, and overly complex strategies are usually destined to fail.

Teams cannot rely on the perfect execution of a double reverse pass on every drive and complicated trick plays, or schemes rarely work in the end. Play books may be long, but the concepts within the playbooks are simplified for players to learn. Moreover, there are methods and naming conventions that help to simplify what are otherwise complicated offensive and defensive concepts. For example, each skill player (quarterback, running back, and wide receiver) is given a number or letter. Further, teams often employ a route tree, with various routes each having a specific number. The Player/Route Number combination can then be communicated rapidly from the coach to the offense (Fig. 7.3).

These principles serve as a method to think about strategy outside the realm of warfare. Establishing strategic principles to guide strategic thinking, planning, and execution—be it on the battlefield or the football field—is critical to strategic success. Further, the principles help define the character of war. While the nature of war remains the same over time, the character of war can change.

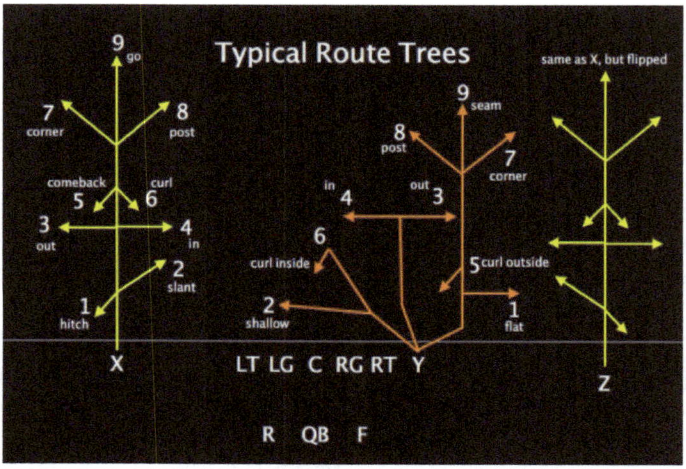

Fig. 7.3 Example of route trees. (*Source* Image by WarroomEagle from CollegeandMagnolia.com)

In addition to the principles of war, the U.S. Military acknowledges three additional principles of joint operations. These include restraint, perseverance, and legitimacy. These three principles came to light as the American Military struggled with the complexity and sophistication of counterinsurgency operations.

Restraint: Restraint is the prevention of unnecessary use of force. This principle came to the forefront as the United States transitioned from large scale combat operations to counterinsurgency operations in both Afghanistan and Iraq. Restraint is important as excessive force can risk the unity of an alliance or coalition, the support of domestic and foreign populations, and degrade the ability of the military to attain national-level objectives. Indeed, each civilian casualty moved the United States closer to defeat in Afghanistan.

At the professional level of football, restraint rarely applies. Players and coaches are paid large sums of money to score on offense and to prevent the other team from scoring on defense. Moreover, players often have performance bonuses within their contracts that encourage them to build up their stats, including touchdowns, rushing, and receiving yards, or sacks. Further, in football, players who don't put forth 100% effort place themselves at risk of injury, as other players on the field are going 100%. There is a smaller feature of restraint when teams build large and seemingly insurmountable leads in a game and insert backup players to prevent injuries to key starting players and to reduce the wear and tear that takes place throughout the season. Other times a team may have clinched a playoff spot and have nothing at stake. The latter often occurs in the last week of the regular season.

At levels below professional football, restraint comes into play in a similar manner, when teams put in second or third-string players once the score gets out of hand. Not only does this prevent injuries to the star players, but also displays good sportsmanship to kids who will never earn a penny playing professional sports.

Perseverance: This principle of joint operations speaks to commitment, and the amount necessary to attain strategic objectives. War is uncertain and unpredictable. Moreover, military operations tend to last longer than expected and can take years if not decades to achieve victory. The reasons for long timelines are legion but include everything from the forces or capabilities available to the actions and reactions of the enemy. From Bunker Hill to Yorktown was six years, Bull Run to Appomattox was four years, as was Pearl Harbor to VE and VJ Days. The Vietnam War lasted approximately 14 years,[11] and the United States kept forces in Afghanistan for twenty years following the attacks of September 11th.

Football teams must persevere through lone and sustained regular seasons. Even further, the end of the season is often just a signal to begin preparing for the next. Players often play hurt and at times through an injury. At the end of a long season, it is often the team with the healthiest players that can carry a team to a championship. More than health, mental perseverance can carry a team through tough times characterized by losing streaks, last-second losses, and the occasional blow-out loss. Football seasons are not decided in one game but rather through the course of a three-to-four-month campaign.

Legitimacy: In addition to physical terrain, the moral and legal high ground is also decisive terrain. In modern warfare, the day-to-day actions of the belligerents are often on display for all the world to see. Smartphones and social media bring the front lines of combat into the homes and lives of people all over the world, not just those in physical proximity to a conflict. Military commanders in modern-day conflict must understand how technology will impact their ability to maintain legal and moral authority over an enemy.

A nation and its military can quickly lose legitimacy through the actions of individuals and units on the battlefield. War crimes and disproportionate responses can reduce legitimacy and thus erode a coalition or support from the local population. War crime incidents such as the massacre of Vietnamese civilians (men, women, and children) at Mai Lai in the Vietnam War, or the rape and murders of an Iraqi young girl and her family in Yusifiah, Iraq by U.S. troops had severe consequences for the individuals involved and for the forces still engaged in the fight.[12] This dynamic is not new as in the Revolutionary War the British continually lost legitimacy with the populations in America through harsh treatment of civilians and the fact that each battlefield victory meant the killing of Americans.[13] Further, as Andrew O'Shaughnessy notes in his book *The Men Who Lost America*, both the British and the Americans committed atrocities, but the British atrocities received the overwhelming press coverage in America.[14] Killing members of the local population is not an effective way to earn their support.

Football teams and players maintain their legitimacy in how they act on the football field, and how a team seeks to gain competitive advantages over other teams. At the individual level, players who develop a reputation as a dirty player can often hold the attention of the ire of referees, thus making a penalty and lost yards more likely throughout a game. There is a long history of so-called dirty players in the NFL including players such as the Jack Tatum, Vontaze Burfict, and Ndamukong Suh, a defensive end who played for multiple teams including the Detroit Lions, the Miami Dolphins, the Los Angeles Rams, and the Tampa Bay Buccaneers. Suh, an all-star player and one of the best at his position had been fined and suspended multiple times for his conduct on the field, diminishing his other accomplishments.

At the team level, when organizations engage in sketchy or illegitimate behavior, they gain a reputation for cheating. This reputation then builds upon itself creating the perspective that every action or event has a malign intent. Moreover, this reputation can cast a dark shadow on the organization's accomplishments in the eyes of the public. Perhaps the best example of this phenomenon is the New England Patriots. During their two-decade reign as the best team in the NFL, the "Spygate" and "Deflategate" controversies distracted from their unprecedented accomplishment on the field.

7.2 Conclusion

Moving from the science to the art of strategy, a strategist must understand when to sacrifice one principle of warfare to achieve another. For example, to achieve surprise on the battlefield may require taking risk in the ability to mass forces. The Doolittle Raid over Tokyo in WWII serves as an example. Taking extreme risk to achieve a level of surprise meant minimal forces applied to the raid. Looking at the art of strategy on the football field, a head coach may do the same when calling for a bootleg with the quarterback. The result is fewer players blocking for the ball carrier, but more yards gained due to achieving surprise on the defense.

The application of principles of war and principles of joint operations is not a zero-sum game. Military leaders and football coaches must select which aspects are most germane when developing their theory of victory. Leaders can emphasize one or multiple principles while also communicating which principle where they may accept risk. The outcomes of wars and football contests are always uncertain, and there is no checklist of which principles to apply and when to apply them. Thus, applying the principles of war is an art and not a science.

Notes

1. Antoine-Henri Jomini, *The Art of War*. (El Paso Norte Press. El Paso TX, 2005).
2. Within this chapter, each of the principles of war and principles of joint operations are defined in *Joint Publication 3–0, Joint Operations* within Appendix A.
3. Michael Lewis, *The Blind Side: Evolution of a Game*. (W. W. Norton & Company. New York, NY, 2007).
4. Super Bowl LIII Box Score: https://www.espn.com/nfl/boxscore/_/gameId/401038115
5. Kathleen J. McInnis, *Goldwater-Nichols at 30: Defense Reform and Issues for Congress* (Washington, DC: Congressional Research Service, May 11, 2016), 7–8.
6. Adam Lazarus, *Super Bowl Monday: From the Persian Gulf to the Shores of West Florida – The New York Giants, the Buffalo Bills, and Super Bowl XXV*. (Lanham, MD: Taylor Trade Publishing, 2011). Kindle.
7. Only the head coach can throw the challenge flag. Although the head-coach will listen to the advice of assistant coaches who may have had a better view of the play, the decision falls on the head coach alone.
8. Michael Lewis, *The Blind Side: Evolution of a Game*. (New York, NY, W. W. Norton & Company, 2007).
9. John Zenor and John Tranchina, "Funny photo boards used to call signals from sideline." *USA Today*. October 10, 2015. Retrieved from: https://www.usatoday.com/story/sports/ncaaf/2015/10/10/funny-photo-boards-used-to-call-signals-from-sideline/73722032/
10. Bettina Renz, *Russia's Military Revival*. (Cambridge: Polity Press, 2018), Kindle.
11. From 1961 to 1975. Of course, to the Vietnamese the war began decades earlier and continued after the Americans left with a war against China. Americans tend to define the timeframes of wars based on American participation.
12. Jim Frederick, *Blackhearts: One Platoon's Descent into Madness in Iraq's Triangle of Death*. (New York, NY: Random House Books, 2010).
13. Andrew Jackson O'Shaughnessy, *The Men Who Lost America: British Leadership, the American Revolution, and the Fate of the Empire*. (New Haven, CT: Yale University Press, 2013).
14. Ibid.

Chapter 8
Application of Joint Doctrine and Joint Functions

He who defends everything defends nothing. – Frederick the Great.

Scripting was a preprepared format, a flexible blueprint that I used to navigate through the turmoil, uncertainty, and stress of competition. – Bill Walsh.

The military perspective of doctrine considers "the fundamental principles by which the military forces or elements thereof guide their actions in support of national objectives."[1] At the tactical level of war, units use doctrine as a method to reduce uncertainty in moments that require the cohesive actions of a team. At the operational and strategic level, decision makers use pre-planned response options and contingency plans to react to various crises around the globe. These response options range from TLAM strikes against an enemy headquarters to the rapid deployment of ground forces into a theater of operations. Contingency plans range from a war plan against a specific adversary to a general plan to provide humanitarian assistance following a natural disaster such as a hurricane or earthquake.

Doctrine is critical to the employment of military forces. Knowing and understanding doctrine allows military leaders at all echelons to act and respond rapidly to unfolding situations. Doctrine provides common terminology, so when a military force is given the task to degrade, disrupt, or destroy an adversary, all units in said military force understand the nature of the task. Moreover, doctrine allows a commander to make assumptions on what higher and adjacent headquarters and units are doing on a battlefield when communications to those elements is not possible. As Trent Hone described in his book *Learning War*, "Doctrine becomes as shared mind-set—a shared context for action—that allows coordinated operations without real-time communication."[2]

Just as military forces have their doctrine, football teams employ their philosophies and game plans at the individual and team level. At the individual level, players at each position learn various techniques or tactics to employ over their opponents. Moreover, there are techniques that apply in various game situations. Lineman, for example, employ run-blocking and pass-blocking techniques dependent upon the play call. Further coaches typically practice and rehearse scripted plays on the first

D. Sukman, *American Football and the American Way of War*,
https://doi.org/10.1007/978-3-031-55345-5_8

couple of drives, the 2-minute drill to conclude a half or game, and trick plays ensuring the team is ready to execute on game day.

At the team level, the offense, defense, and special teams employ a scripted set of plays. For example, the offense may employ a series of set plays during the execution of a 2-minute drill, when time is limited, and scoring is paramount. Conversely, a defense may execute their pre-scripted formations in a goal-line situation when the offense only needs 1 or 2 yards, or in a 3rd down and long situation, when the offense needs to gain 10 or more yards to gain a first down. At the highest levels of the game, nearly all scenarios have a scripted play design that is rehearsed repeatedly in practice. This is no different from units rehearsing their battle drills before engaging on a mission and conducting actions on the objective. Indeed, military units will spend countless hours, days, and weeks preparing for missions and specific combat situations.

Of course, with the advantages of doctrine come its limitations. Doctrine may be descriptive and provide sound methods for solving tactical and operational problems, but it is not prescriptive. Commanders at all levels are free to deviate from it in the same way quarterbacks call audibles at the line of scrimmage. Further, doctrine does not reach outside of specific military organizations. Marines do not follow army doctrine, and State Department employees do not abide by Joint Doctrine. So, while doctrine does provide a common language, it is not a panacea to the friction all forces and organizations encounter in warfare.

8.1 The Joint Functions

At the operational level of war, joint doctrine breaks war down into seven different bins or "joint functions" as a mental construct for all the various capabilities and activities on the battlefield. These seven functions include command and control or C2, movement and maneuver, fires, sustainment, protection, intelligence, and information. Understanding each of these functions is useful to guide operational planning by framing the capabilities available to a commander. Moreover, planners use the joint functions to describe how their respective commanders will use various capabilities and combat platforms on the battlefield. Indeed, the joint function framework helps planners account for and address all aspects of an operation.

Just as the military organizes and bins capabilities and activities, so do football organizations at each level from Pop Warner to the National Football League.

Command and Control: Simply put command and control is about authority and direction. In the military, at the tactical level of war, command and control is often simply communicating and directing one echelon to another in a top-down manner. Squads report to platoons, platoons report to companies, and companies report to battalions. It is at the operational and strategic levels of war where command and control become complicated. Operational and strategic level organizations often work with parallel commands and often provide support to other commands who are competing for the same resources. There are four principles to command and control which include simplicity span of control, unit integrity, and interoperability.

Simplicity means that there is an unambiguous chain of command and that command relationships between units are specified and understood. Moreover, simplicity means that the authorities and responsibilities of command at each echelon are clearly defined. A tenant of simplicity is unity of command means the concept that one person is in charge. All activities conform to the desires of a single commander who holds unquestioned authority over the organization. This is paramount in military operations; particularly when activities occur in the land, air, and maritime domain where multiple services (Army, Navy, Air Force, Space Force) participate provide forces. The unity of command principle occurs in multiple phases of the football game. In the huddle, for example, the quarterback is in charge, and the players answer to him. For the team, the head coach is in charge and is the final decision authority in game planning and play calling. Ultimately, when it comes to the running of a professional team, the owner is the one with singular authority to make all the decisions of a franchise.

Span of control speaks to several factors but is best understood by the point that human beings can manage only so many things at one time. There is a limit to the number of subordinates, activities, force capabilities, size of an operational area, or range of weapons systems that one person or one headquarters can manage. To manage span of control, leaders establish subordinate commands and limit the number of direct reporting subordinate units. Properly managing span of control helps leads make better and more informed decisions while increasing their situational awareness. In the U.S. military, perhaps the seminal document in guiding span of control is the Unified Command Plan (UCP). This document divides the world into geographic regions and appoints a command for each region such as U.S. Central Command for the Middle East, U.S. European Command for Europe, and U.S. Africa Command for Africa.[3]

Every coach has a limit to the number of players and assistant coaches they can supervise. The head coach of a football team does not have the time or capacity to run every drill for every position in practice. Teams have offensive and defensive coordinators, a special teams coach, and at higher levels of football, coaches for each position including offensive and defensive lines, quarterback coaches, wide receivers, and defensive backs. Further, teams often employ strength coaches and athletic trainers to manage the physical fitness of the team. The quality of the coaching staff is a direct reflection of the head coach as much as the performance of the players on the field.

Unit integrity is the idea that forces should remain organized as they were designed, and in the way they trained before the mission. Although commanders often have the authority to reorganize units and assign them to subordinate commanders as they see fit, maintaining unit integrity is one of the preferred methods to fight as you are trained. Keeping unit integrity supports mutual trust between various headquarters and reinforces already established roles and responsibilities.

Football teams that stay together have a competitive advantage over teams that are continually shuffling personnel throughout the season. Further, stabilization at key positions such as quarterback is often decisive to longer term success. Even more important may be stability at the head coaching position with various assistant coaches maintaining their jobs as well. In the era of free agency and in a sport with a high rate of injury, it is increasingly difficult to keep teams intact for more than a few seasons. Indeed, the era of dynasties staying together over the span of a decade, such as the 1970s Steelers or 1980s 49ers is long over. Further, the best assistant coaches are continually on the prowl seeking their own head coaching gig. But there are some examples of continuity and team integrity since the turn of the century, including the quarterback and head coaching of the New England Patriots, the Pittsburg Steelers, and the Green Bay Packers to name a few.

Interoperability means that units should do their best to work for or with other units with similar equipment, doctrine, and policy among other elements. For example, various military units would want their radios and other communications equipment to talk to each other, or to use similar doctrine and language to ensure orders are followed within their intent. Members of the NATO alliance for example use the same caliber ammunition as a method of interoperability. Moreover, NATO has its own doctrine independent of member nation doctrine to ensure that when NATO conducts a military operation all contributing forces have a common understanding and use common terminology.

On the gridiron interoperability is a bit different. Coaches must find the right talent and players who fit into their systems or adjust their systems to match the talent they inherit on the team. Not every player can fit neatly into a 3–4 or a 4–3 defense. Not every quarterback can operate the run-pass option (RPO). The key is to ensure players are not asked to execute a system for which their talent doesn't match. A terrific example of a team adjusting to its level of talent is the Army Football team who in 1984 adopted the wishbone offense with the understanding that their smaller and generally slower players cannot match up to larger Division-1 schools. By adopting the wishbone, Army slows the game down with longer drives keeping high-powered offenses of Notre Dame and other powerhouses off the field.[4]

The last aspect to discuss in command and control is the idea of mission command. Mission command is an old concept that dates to the German or Prussian concept of *Auftragstaktik*. The fundamental behind this idea is that control from higher headquarters should be based upon trust and loose direction rather than tight and detailed instructions. Higher level commanders provide a mission and intent and largely leave the details of accomplishing the mission for development and execution by the subordinate commanders. Mission command is predicated upon trust up and down the chain of command.

Some football teams leave the running of all football operations to the head coach, leaving them free of administrative responsibilities. Further, some head coaches maintain control of personnel decisions as the owner or senior management of the organization trusts the head coach to draft and sign the right players. Some coaches in turn will either control the play calling or place that responsibility with their offensive and defensive coordinators. Even further when there is an experienced quarterback the coaching staff will have the trust and confidence to allow the quarterback to call his plays, and to audible whenever he feels the situation requires it.

If you can't get command and control right, there is probably little else you will get right. An organization with unclear command and control relationships is detrimental to the mission. The value of having a clear chain of command with effective leaders who understand the limits of their authority ensures that subordinates unambiguously know who is in charge is essential to guiding all members of an organization towards common objectives.

Movement and Maneuver: This joint function focuses on the positioning of joint forces and capabilities to positions of advantage before, during, and following combat operations. The movement of forces can occur in peacetime and war. An example of movement in peacetime is the securing of basing rights in a foreign nation and then placing forces on a rotational or permanent basis. This could be a forward operating base for the army, a port for the navy, or an airfield for the Air Force. In addition to the movement of large-scale capabilities in peacetime, commanders must maneuver tactical and operational forces while engaged with the enemy. Movement and maneuver is the process by which commanders set the terms and conditions of battle in time and space, to decline battle, or to exploit an existing situation.

Football at its core is a contest of movement and maneuver. Throughout a game, each team seeks to put their players in a position to advance the ball on the field while on offense, or to prevent the opponent from moving the ball forward. Often, games become focused on field position as teams seek to keep their opponent positioned deep in their own territory. Moreover, drives that begin in opposition territory are more likely to result in points by either a field goal or a touchdown.

Fires: Military capabilities aligned to the fires joint function include targeting, artillery, strategic attack systems, and missiles to name a few. Fires are how military forces create lethal and non-lethal effects on the battlefield. Typically, military units employ fires to deliver lethal effects which result in destruction, degradation, or disruption of enemy forces and systems. Fires that employ non-lethal effects such as electronic attacks (in the electromagnetic spectrum), or network attacks in cyberspace tend to limit collateral damage and reduce risk to civilians and other non-combatants. Military commanders balance the advantages and disadvantages of each when thinking about and deciding which fires method to employ.

On the football field, teams must decide how they will attack and defend against their opponents. Fires cannot be divided into lethal and non-lethal categories on the gridiron, but coaches must make decisions on when to call plays that seek to attain a few yards against when they assume more risk in attempts to advance the ball down the field for larger gains. Multiple factors go into this decision making which can include the current score, the number of yards required for a first down, and the time remaining in a half or in a game.

A second example of fires within a football game is the kicking aspect. Teams strive to score touchdowns but when their drive stalls in opposition territory an accurate kicker who can make field goals from long-range distance is worth their weight in gold. Often, kickers are the decisive factor in close games with the last second field goal attempt determining the winner and loser of a game. As of 2023, five Super Bowls ended with a last-second attempt to include Jim O'Brien's Super

Fig. 8.1 Scott Norwood's field goal goes wide right in super bowl XXV. The giants win 20–19. (*Source* AP Photo/Phil Sandlin)

Bowl V field goal and Adam Vinatieri's last second kicks to win Super Bowls XXXVI and XXXVIII. Scott Norwood's last-second attempt that went wide to the right was the deciding moment of Super Bowl XXV. Super Bowl LVII ended with the Chiefs' Harrison Butker booting a 27-yard field goal to win the game (Fig. 8.1).

Just as there are offensive fires in a football game, there are defensive fires in the form of kickoffs and punts. Again, accuracy and range are critical to placing the opponent in a position of disadvantage deep in their territory. A team starting a drive from inside its own 10-yard line is less likely to score than starting at mid-field.

Sustainment: In his book *Moving Mountains*, retired General William Pagonis tells us every army in the world today does the following three things; "they eat, they move, and they shoot."[5] This translates into a massive necessity for food and water, fuel, and ammunition. Providing sufficient logistics and personnel support is the primary purpose of the sustainment joint function. At the operational and strategic level, proper sustainment ensures that military forces have freedom of action and can extend operational reach deep into an enemy's area of operation. At the tactical level, proper sustainment is essential for maintaining morale among the military serving in lands far away from home. Indeed, activities in the sustainment joint function range from supply and maintenance to religious ministry and postal services.

At the lower levels of football, the logistics of the game often relies on the coaches, parents, and other volunteers in the league. Buses to transport the team to the next town, uniform issue at the start of each season, and reserving set times for the team to work in the gym, use a film room, or have a practice field to themselves are considerations for under-appreciated men and women behind the scenes. Other aspects of sustainment include medical support, which is especially critical to the violent game of football. A quick glance at the sidelines of any game reveals athletic trainers, medics, doctors, as well as ambulances at the ready to take an injured player directly from the field to the emergency room. Indeed, without these efforts, the events on game day simply do not occur. Organizations that don't apply the right people and resources to sustainment and logistics have a recipe for failure.

Just as operational planners generally leave the sustainment and logistics planning to professionals in the field, so do professional football teams who rely on the administrative and logistical support off the field. Coaches and players focus their efforts on planning for and playing in their respective games, while the coordination for transportation, equipment, and even pre-game meals is the domain of the front office. In each case, there must be dialogue between the planners and the logisticians or the coaches and the front office to ensure the right needs are met at the right time. Further, there is nothing so dangerous in the military as maneuver forces (tanks, trucks, troops) getting too far away from their sources of supply.

Protection: During combat operations, military forces take actions that help in preserving future combat potential. This ranges from active measures such as air defense systems (e.g., Patriot Missiles and SM-3 s and SM-6 s) to passive physical protection such as concrete walls and barriers to prevent and guard against medical and health threats through vaccinations and proper sanitation. Protection is useful not only in preserving combat power in the air, maritime, and land domains, but also applies equally to space-based assets and cyberspace capabilities.

Protection in football seems obvious at the tactical level. Teams build offensive lines to protect their quarterback. However, there is a deeper essence of protection when it comes to football. Certainly, on the playing field, the ability of linemen to protect the quarterback is built into the player selection and passing protection schemes. On the sidelines, coaches often use deception and other types of protection measures to ensure opposing teams can't determine the play calls. From covering their mouths when they speak to having multiple play signalers on the sideline, there is thought and process to protect game plans and play calls. Off the field, teams employ athletic trainers and strength coaches to preserve their team by preventing and mitigating injuries to their players.

A common thread in protection since the COVID-19 pandemic hit the United States is the competitive advantage a healthy force and team gains when its members are vaccinated. For example, early in the pandemic and before the development of COVID-19 vaccines the U.S. Navy pulled an aircraft carrier offline due to an outbreak on the ship.[6] Medical capabilities are critical to the health of a military force and serve as a force multiplier.

In the NFL and college, players and coaches missed multiple games due to positive tests. In certain instances, key players could not play, resulting in losses and a failure of the team to make the post season. Strength trainers and physical trainers are a vital part of a team's coaching staff. Moreover, teams maintain a cadre of doctors and medical specialists to assist in the rehabilitation of players who sustain injuries. A roster of healthy players is a competitive advantage in football.

Intelligence: In military parlance, intelligence supports how a commander sees and understands the operational environment. Indeed, it is intelligence that tells the commander what the enemy is doing, what they are capable is doing, and what they are expected to do in the future. Intelligence informs the decisions of the commander. Before every decision, a commander looks at what he or she understands about the enemy, and what he or she understands about themselves. Commonly referred to as Commander's Critical Information Requirements (CCIR), this information is broken into priority intelligence requirements (PIR) which focuses on the enemy and the environment, and friendly forces information requirements (FFIR) which focuses on friendly forces and supporting capabilities.

To collect information, military commanders employ a variety of intelligence, reconnaissance, and surveillance (ISR) capabilities across the battlefield. ISR capabilities range from spy satellites in space watching the enemy to electronic intelligence platforms listening to the enemy, to human spies talking to and interacting with the enemy. These capabilities assist commanders and their staffs in identifying enemy weaknesses, or indicators of what course of action an adversary will execute.

During a season, football teams at every level seek to gain an understanding of their opponents and their own team. With the former, football teams send scouts to opposing teams' games and use film to study the habits and tells of opposing teams. Watching other teams on film allows a coach to identify weak points in an opposing defense or offensive line to exploit throughout a game. With the latter, head coaches continually monitor the health, well-being, and talent of their team to inform their respective game plans. Having this information ready allows coaches to develop advantageous personnel matchups in the game. An easy example of this type of thinking is that the best coaches will not demand more from their players than they can give. They design systems and game plans that match their available talent against each opposing team.

Key to the design of a game plan is the work of football scouts. Scouts are football's version of ISR. It is the scout's job to watch the other football teams and to look for patterns in their play calling and formations and find tells in the way players set themselves before a play. Further, scout determine the strengths and weaknesses of an upcoming opponent, identifying what areas of the field or what players the team can focus its efforts on.[7] This allows the head coach and his or her assistants to develop a game plan, and to counter or render ineffective the thing that the opponent does best.[8]

The tricky aspect of intelligence is what level of confidence one should place upon it. Two of the theorists discussed earlier in the book, Clausewitz and Sun Tzu came to opposite conclusions on the value of intelligence.

To Clausewitz, the uncertainty of war and the combined aspects of fog, friction, and chance made intelligence suspect at best. Clausewitz observed that intelligence reports are often false and contradictory. Moreover, Clausewitz understood that when it comes to human intelligence (HUMINT), most people tend to exaggerate the bad news, and leaders tend to place more faith in bad news than good.[9] An easy example of the exaggeration of bad news and the overestimation of enemy capabilities lies in the early days of the U.S. Civil War. Indeed, one of the reasons General George C.

McClennan was hesitant to move his army and attack the confederates from 1861–1862 was his flawed belief that the enemy outnumbered his forces by a factor of 2–1.[10] Intelligence assists commanders in making decisions, but the commanders remain responsible for evaluating the intelligence and for the decisions they make.

Opposing Clausewitz's disbelief in the power of intelligence is Sun Tzu who framed intelligence as the most important aspect of war. Sun Tzu's advice to know thy enemy and know thyself emphasizes the value of collecting information on the enemy, and analyzing that information, turning it into actionable intelligence for the commander.[11]

Military commanders and football coaches alike should never place blind faith in the information or intelligence they receive. Rather, they should question sources and think critically about intelligence before making decisions that will impact the course of a game or the outcome of a battle. Adversaries may know about collection activities, and use that to gain an advantage. Further, they should consider the previously discussed aspect of reciprocity, or consider the possibility of deliberate deception on the part of the opponent. This ability is gained over time through a combination of practice and experience.

Information: This function "entails the management and application of information and its deliberate integration with other joint functions to influence relevant actor perceptions, behaviors, and actions or inactions."[12] Information related operations and activities include key leader engagement, public affairs, and military information support operations (formally known as psychological operations or PSYOPS). With the advent of the internet and other forms of instantaneous global communication, information in war is more important than ever. The ability to use information to one's advantage can turn tactical losses into operational or strategic victory. Hezbollah's use of information was central to its claims of victory following decisive battlefield defeats in Lebanon in 2006. Despite the massive destruction inflicted by the Israeli Defense Forces (IDF), Hezbollah was able to claim the mantle of leadership throughout all of Lebanon.

The information function should be familiar to anyone who played or observed the game of football. At the lower echelons of the game, teams often develop methods to intimidate their opponents. Methods vary from the way they conduct themselves in pre-game warm-ups to the interaction with a friendly home crowd. At the higher levels of college and professional football, teams often use various methods of information to motivate their own players. There are countless examples of players taking soundbites of an opponent disrespecting them in the press to up the intensity of their own game. One of the more famous examples in the NFL occurred just before Super Bowl XXVI when the Bills' defensive line coach Chuck Dickerson publicly mocked the Redskin's offensive line. The Redskins used this information as a motivation tool and routed the Bills, handing Buffalo its second of four consecutive Super Bowl losses. Coach Dickerson was looking for a new job three days later.[13] One of the easy ways to motivate an opponent, is to have disrespectful sound bites in the media.

8.2 Conclusion

Binning capabilities within the seven joint functions is a useful method for planners to account for all the needs of the forces they wish to employ. Leaders, whether on the battlefield or the gridiron should account for all the capabilities available to their organization and in turn, align them towards common goals or objectives.

Notes

1. *Joint Publication 1–02 Department of Defense Dictionary of Military and Associated Terms.* February 2016. (Washington, DC: The Joint Staff). Pg 71.
2. Trent Hone, *Learning War: The Evolution of Fighting Doctrine in the U.S. Navy, 1898–1945.* (Annapolis, MD: Naval Institute Press, 2018), 12.
3. As of 2023, the U.S. military has 11 combatant commands, 7 of which have geographic areas of responsibility, and 4 with transregional responsibilities. 1) European Command, 2) Africa Command, 3) Northern Command, 4) Indo-Pacific Command, 5) Southern Command, 6) Central Command, 7) Space Command, 8) Special Operations Command, 9) Transportation Command, 10) Cyber Command 11) Strategic Command
4. John Feinstein, *A Civil War – Army vs. Navy: A Year Inside College Football's Purest Rivalry.* (Boston, MA: Little, Brown and Company, 1996). 17.
5. William G. Pagonis and Jeffrey L. Cruikshank. *Moving Mountains: Lessons in Leadership and Logistics from the Gulf War.* (Boston: Harvard Business Review Press, 1992).
6. Dan De Luce, "Aircraft carrier failed to enforce social distancing, allowing Covid to spread: Pentagon watchdog." NBC News. 8 February 2021. Retrieved from: https://www.nbcnews.com/news/us-news/aircraft-carrier-failed-enforce-social-distancing-allowing-covid-spread-pentagon-n1257023
7. Steve Belichick, *Football Scouting Methods.* (New York: The Ronald Press Company, 1962). Kindle.
8. Ibid.
9. Clausewitz, Carl von. *On War*. Eds. and trans. Michael Howard and Peter Paret. (Princeton, NJ: Princeton University Press, 1976). 117–118.
10. Stephen W. Sears, *Landscape Turned Red: The Battle of Antietam.* (New York, NY: Mariner Books, 2003) Kindle.
11. *Sun Tzu: The Art of War.* Trans. Samuel B. Griffith. (Oxford University, 1971). 84.
12. Joint Publication 3–0. *Joint Campaigns and Operations.* (Washington D.C.: The Joint Staff, 2022).
13. *Desert News.* "Bills Fire Defensive Coach. (30 January 1992). Retrieved from: https://www.deseret.com/1992/1/30/18964962/bills-fire-defensive-coach

Chapter 9
Design, Planning, and Problem Solving

> Alice asked the Cheshire Cat, who was sitting in a tree, "What road do I take?" The cat asked, "Where do you want to go?" "I don't know," Alice answered. "Then," said the cat, "it really doesn't matter, does it? -Lewis Carroll (Alice in Wonderland).
>
> Plans are useless, but planning is indispensable. -Dwight D. Eisenhower.
>
> Pressure is something you feel when you don't know what you're doing – Chuck Noll

Distinct from the rigid step-by-step approach of the joint planning process and service equivalents, design is a method to help planners and commanders think about and identify the problem in front of them before they begin developing a solution. Design in its most basic form is describing the current environment, the desired environment, and then figuring out the problem, or what is preventing a command from getting from the current conditions to the desired conditions. What is useful about design is that it allows planners the time and space to think critically about the problems they face. While design is like and complementary to planning, the two are not the same.

Design is more of a methodology that frames the environment and helps a commander and staff gain an understanding of what they are facing. While planning is more of a step-by-step approach, design is an art. Commanders that use design properly will gain an understanding of the problem they face which enables them to use the science of planning to solve said problem. How a force will overcome this problem is expressed through an operational approach, which uses and depicts the elements of operational design.

What makes design and the identification of the problem to overcome important is the time that a planning staff takes the most risk. If the staff and the commander are solving the wrong problem, subsequent planning and execution are fruitless endeavors no matter how hard or long people work. More than identifying the problem, planners and leaders should consider that the first stab is often an immature understanding of the problem. Design is iterative and dynamic, not linear, and static. The problem can evolve and take new forms as planning and execution go into action.

The 2020 version of *Joint Publication 5–0: Joint Planning* lists thirteen elements of operational design.[1] While not every plan or operation needs to address each

D. Sukman, *American Football and the American Way of War*,
https://doi.org/10.1007/978-3-031-55345-5_9

element, they are useful to provide a construct or framework for a plan. Naturally, as every chapter in this book has so far presented, these elements play out on the gridiron just as they do on the battlefield.

9.1 Objective

U.S. joint doctrine describes the objective as the "most single important element of operational design….it is why the mission is being conducted and should be determined first." Understanding the objective of a military campaign or mission is often determined by examining and analyzing strategic guidance. Further, military leaders and planners must participate in a dialogue with civilian policymakers to better understand the purpose of what they plan to accomplish. Moreover, as geopolitics and war are dynamic environments (not static), this guidance continually changes. Guidance can come in the form of documents such as the *National Security Strategy*, in speeches by the President, and yes, since 2017 strategic level guidance can be found on a Twitter feed or other forms of social media. An important aspect in determining an objective is that objectives must be decisive yet attainable. Long term occupations and governance of foreign nations with the objective of establishing a fully functioning Jeffersonian Democracy, for example, may not be an attainable objective (Fig. 9.1).

Critical to using design is a continuing dialog between commanders and senior leaders at each echelon. Subordinate commanders must understand their higher direction and guidance which helps form a clear picture of where they want to be and

Fig. 9.1 Alice in Wonderland's The Cheshire cat. Image from wikimedia commons

what they want the environment to look like at the end of an operation or campaign. Every echelon of command through each level of war (military commanders, football team owners, coaches, and players) must participate in this dialogue to ensure leaders are working towards the same operational and strategic objectives with a mutual agreement on what winning looks like, and what it takes to get there.

Football organizations should identify what the objectives or goals of the organization are, and what the conditions are when those goals are realized. In doing so, like military planners, they can identify what is standing in their way of success and develop an appropriate method to overcome these obstacles. Owners, general managers, and athletic directors should set the expectations and definitions of success for coaches so that they can implement the best plan for the team in the timeframe they are provided. In terms of being decisive and attainable, leaders within football organizations should understand that a championship or other specific goals may not be attainable each season.

Military End State: In the famous tale *Alice in Wonderland*, the Cheshire Cat informs Alice that "if you don't know where you are going, any road will take you there.[2] Identifying what you want the end of an operation or mission to look like is important to help think about where you are now, and what is preventing you or standing in the way of attaining the end states. For military planning, the military end state is the point where the military element of national power is no longer required. It is important to note that the U.S. government may still be involved in using other instruments of national power (Diplomacy, Economic, Information) and other agencies of government. Military organizations look to senior officials and strategic documents to determine what their end states should look like.

There is little equivalent of a military end state in football. There are no conditions where the team or the players are no longer necessary.

Center of Gravity: It was Karl von Clausewitz who originally conceptualized centers of gravity. But is in design where the idea comes to fruition. Military planners look to identify friendly and enemy centers of gravity at the operational and strategic levels. Once identified, planners then seek out the critical requirements, capabilities, and vulnerabilities. Moreover, identification of all these aspects of centers of gravity allows commanders and staff to take a direct approach (attack the COG) or an indirect approach (attack the critical requirements and capabilities). This is the area where Clausewitz and Jomini come together.

Football coaches and scouts continually look for weak or vulnerable areas of an opponent to attack. While the center of gravity might be the opposing quarterback, a team may go after a weak offensive tackle or another lineman to find their way to the quarterback. Indirectly, the best way to negate the talents of an opposing quarterback may be to completely cover all his receivers, thus forcing the opponent to rely on a less reliable running attack.

Subordinate to a Center of Gravity is the idea of critical requirements, critical capabilities, and critical vulnerabilities. Military commanders and staffs use these concepts to analyze the center of gravity of their enemy and themselves. Further, a solid center of gravity analysis that accounts for these elements has the advantage of forcing commanders and their respective staff to think and reflect upon adversary and friendly strengths and weaknesses. This analysis is useful even if the identification of the COG is off, or the COG shifts during the unfolding of the operation.

A critical capability is "a means that is considered a crucial enabler for a center of gravity to function." Military examples of critical capabilities might include communications between leaders or support of elements of the local population (should a regime or leader be the identified COG). Football critical capabilities might be the offensive or defensive line or the combined abilities of the assistant coaches.

A critical requirement is an essential condition, resource, and means for a critical capability to be fully operational. Examples of critical requirements can include the attainment of air and maritime superiority as a condition. Fuel, electricity, and ammunition can serve as resources, as can various weapons systems including air defense, or carrier strike groups. Critical requirements in football might be continual advantageous field position, a consistent and accurate placekicker, or fast receivers who can generate open space for the quarterback to throw the ball.

Critical vulnerabilities are an aspect of a critical requirement that is vulnerable to direct or indirect attacks. On the battlefield, a critical vulnerability may be the extended lines of communication. Other examples could be insecure websites or a mass formation of forces in open terrain. A football team's critical vulnerability may be a weak offensive tackle or a notoriously slow defensive secondary that is unable to cover speedy receivers. Conversely, military planners and coaches should analyze their team or force and identify their respective vulnerabilities or weaknesses. This analysis allows for the protection or mitigation of weaknesses. Planners and coaches seek out the vulnerabilities of their opponents and exploit them while simultaneously protecting their own in their quest for victory.

Direct and Indirect Approach: Commanders must decide on the way they will attack an enemy's center of gravity. This element of operational design combines Clausewitz's center of gravity (COG) concept with Liddell Hart's indirect approach. A direct approach attacks the enemy COG by applying combat power directly against it. The indirect approach takes a different route and gets to the COG by attacking the critical capabilities, critical requirements, and critical vulnerabilities to eventually get to the enemy COG. For example, a direct approach might seek to eliminate a national leader at the onset of the war, however, this decapitation strategy may not be feasible as the national leader is well protected. The indirect approach might seek to starve the national leader of allies and partners or to erode the support of said leader's military and internal security forces.

A football team may seek to attack a quarterback directly through blitzes and other defensive rushing schemes in a direct manner. However, teams often invest in offensive lines to protect their quarterbacks, leading to an indirect approach of blanketing a quarterback's receivers. The term coverage sack comes from a situation where a quarterback simply has no viable receivers to throw to, leading to the inevitable breakthrough of the defensive rushers.

In certain aspects, when football coaches develop a game plan, they must choose to either attack the opposing team's strengths or avoid them. This may mean passing the ball to secondary receivers covered by the second or third-best defensive back or to attack the opposing team's best player, with the underlying thought that their strength is superior to the adversary's strength.

Effects: An effect is a physical or behavioral state of a system that results from an action or set of actions. The best way to think about an effect is that it is a condition, or series of conditions that a military force seeks to attain to accomplish an objective. More experienced planners and staffs will not only identify effects but also undesired effects that can inhibit progress. An example of an effect could be the enemy (actor) can no longer conduct offensive operations (a measurable behavior).

Football teams also create effects on the gridiron. While an objective for the defense might be to limit the number of points the opposition scores, the effect could be the opposing running back (actor) will not surpass 100 yards rushing (a measurable behavior).

Culmination: Simply put, culmination is the point in time or space where an operation can no longer maintain momentum. For forces attacking an enemy, it means that the attack can no longer continue and that units should transition to a defensive posture as an operational pause or to consolidate and resupply before continuing the attack. For forces fighting a defensive battle, culmination occurs when they can no longer defend a position. More than recognizing culmination in friendly forces, commanders should seek to put their adversary in a position where they can either no longer conduct an attack or defend a position.

At the operational level, attacking forces typically achieve culmination when they outpace their support. When maneuver units move faster and farther than those that provide them with fuel, ammunition, water, and food, they must consolidate and wait for resupply lest they be cut off and isolated behind enemy lines. Tactical level shifts can disrupt an operational level of plan, in that they may require reshuffling of the fires, logistics, and intelligence support systems. The next order of effects is that units who were expecting this support may not receive it, putting the plan and operation at risk.

At the strategic level, a nation can culminate through the erosion of national will and/or legitimacy. Excessive casualties, for example, can lead to a decline in national support, as can the long-term occupation of a foreign nation combined with a steady dose of bad news. The United States is no stranger to strategic culmination, as the population grew weary and uneasy during the Vietnam War, and later in the wars in Iraq and Afghanistan.

Culmination occurs in football at the tactical and operational levels. Tactically, during a game, each drive has a culminating point. There are times when an offense can sustain a drive deep into the opponent's territory, only to lose momentum and settle for a field goal. Other times, a defense can hold back a powerful offense for part of the game, only to have the floodgates open in the final quarter. At the operational level, there is typically a set amount of time a head coach must bring a sustained level of success. Some coaches collaborating with general managers can find it instantly, making a playoff run early in their tenure, followed by a steady decline in wins over successive years. This latter aspect of culmination often coincides with the aging of athletes on a team whose physical prowess tends to decline after multiple seasons of violent collisions.

Lines of Operation: Lines of operation or LOOs are another of Jomini's ideas. LOOs are a way to progress through a military operation by using physical terrain to measure progress. A famous example of is Douglas MacArthur's island-hopping campaign in the Pacific during World War II. MacArthur's forces moved from one island to the next in a steady advance closer and closer to mainland Japan. More recently, the U.S. military in the 2003 invasion of Iraq used lines of operation in their advance from the Iraq-Kuwait border to Baghdad. U.S. forces sequentially captured a series of cities or physical terrain.

Throughout a football game, each drive could be considered its own line of operation. Teams steadily advance down the field toward the endzone. Progress is measured in yards and first downs. Field position, while not the defining element of wins and losses, has a substantial impact on the outcome of a game.

Lines of Effort: Opposed to lines of operation which are physical, lines of effort are temporal. Lines of effort became the popular method of planning throughout the counterinsurgency campaigns in Iraq and Afghanistan as the military was no longer seeking to capture and occupy physical terrain from the enemy. Popular lines of effort included governance, security, information, and development.

A way to think about lines of effort vis-à-vis football is through the several ways a team seeks to win either a game or run the organization from the strategic perspective. At the game or tactical level, a coach may have three lines of effort that include offense, defense, and special teams. At the operational and strategic level looking beyond a single game or season's lines of effort for a team's campaign could include player acquisition, marketing, and ticket sales.

Decisive Points: A decisive point is a point in space or time where one gains a marked advantage over the adversary. Commanders and staffs try to identify decisive points that are related to centers of gravity, specifically the critical capabilities, requirements, and vulnerabilities. Popular examples of decisive points are the attainment of air or maritime supremacy over an enemy, having a preponderance of combat power in a theater of operations, or receiving overflight permissions from an ally or partner.

Decisive points can occur in a game, throughout a season, or from the strategic perspective throughout a coach's tenure with a team. In-game decisive points can include touchdowns and field goals, sacks, interceptions, and other turnovers. In season decisive points could be each victory and longer-term decisive points could include key draft picks, key free-agency signings, or the hiring of sought-after assistant coaches and coordinators.

Operational Reach: This is simply the distance and duration across which a unit can successfully employ military capabilities. Everything has a limit, including the physical distance a weapon can shoot, the distance tanks can travel before running out of fuel, or the time a crew can spend submerged in a submarine. Sometimes operational reach is constrained by geography, other times by the limits of machine or human endurance. Commanders must know and understand the limits of their force, and what they can achieve before they assume unacceptable risk.

In the contemporary environment, the expanded Russian invasion of Ukraine in March of 2022 exposed the limits of Russian operational reach. Indeed, Russian land forces were limited in how far they could travel, and how much territory they could occupy and secure. Further, Ukrainian forces specifically targeted Russian sustainment, limiting the ability of Russian forces to resupply units attempting to advance deep into Ukrainian territory. The result of this limited operational reach was the mauling of the Russian army.

Within a football game, coaches should understand the reach of their team throughout a game. While some teams have the capability (talent) to execute long passes and other plays that result in chunks of yards, other teams have a smaller reach and move the ball a few yards at a time, taking shots at longer gains on a less frequent basis. Limiting operational reach in this respect is the arm strength and accuracy of the quarterback. Football teams also have an operational reach with their special teams. Kickers and punters alike possess a limit to how far they can kick the ball. Coaches understand the limits of their reach and make decisions based on this information accordingly.

Arranging Operations: In a military operation a commander must decide how to arrange his or her forces. The term "arranging" speaks to simultaneous or sequential actions to achieve the desired objectives and end-state. It is the arrangement of forces that will determine the tempo or pace of the operation. To plan the arrangement of forces commanders also consider phasing, branches and sequels, and operational pauses.

The idea of simultaneity is simple, it is about attacking multiple formations of an adversary and can occur over dispersed geographical spaces. It forces an enemy to prioritize and decide where they will focus their warfighting capabilities. Simultaneity can create confusion and chaos in an adversary's ranks.

Phases of an operation are typically sequential; however, some aspects or actions may overlap. Each phase in a campaign plan typically represents a distinct major operation or series of actions. A team may break down a single game into four phases (1 quarter = 1 phase) while a head coach may look to break down the season into any number of phases based on the schedule. Further, a head coach, general manager, or owner may look to a longer-term plan with build in phases to build up a team over the course of several seasons and off-seasons.

In military operations, transitions between phases often come with an increased risk to the force. When moving from one phase to another, the command and control arrangements might change, as may the authorities and rules of engagement. Moreover, what force or unit comprises the main or supporting effort can shift from one phase to another. Because of these shifts, military commanders must pay close attention to operations during transitions. Like war, the most dangerous or risky period of a football game is the transition from one phase to the next. The transition from defense to offense often comes in the form of a punt or kick-off return. Indeed, muffed punts and fumbled kick-off returns can dramatically shift the field position and change the momentum and outcome of the contest. Blocked punts when transitioning from offense to defense offer similar levels of risk.

Identifying branches and sequels is one of the most important aspects of military planning. Branches of a plan answer the question "what if?" while sequels answer the question "what next?" If history is our guide, then sequels are the harder of the two to conceptualize while planning and developing a strategy or operation. Napoleon for example failed to ask himself "what next" when his forces invaded Russia and occupied Moscow. What came next was a long retreat that dwindled his once Grande Armée into a shell of its former self. In the modern era, the United States Army had little clue about what to do once it occupied Baghdad and other major cities following the 2003 invasion of Iraq. Not planning for the next phase was a key contributor to the rise of an insurgency and civil war within Iraq, one that the U.S. Army was ill-prepared for.

At the end of each season, every football organization must ask itself "what do we do next." In youth, high school, and college players will move on to the next level, transfer to another school, or simply quit the game. Coaches must anticipate the rapid turnover of players and think about how they will recruit and develop the next cohort of players. At the professional level, similar scenarios play out as older players retire, others sign a contract with another team during free agency, while others simply age a year and return a little bit slower and less effective than in the prime of their careers. Moreover, assistant coaches may look for head coach positions within the league or at another level of the game. From the front office perspective, the same dynamic can play out with the hiring of a new general manager and head coach. Regardless, teams must consider how to make their team better for the next season. Indeed, even championship teams must find ways to improve in the offseason as other teams in the league are aiming to take the mantle the following season.

Branches provide an element of flexibility to a plan. Should events unfold in unexpected ways there should be some forethought in how to react to prevent complete failure. For example, if the enemy appears at an unanticipated place on the battlefield, commanders and staffs should have the flexibility in the plan to change course and react accordingly. Other times an allied nation might deny access or basing on their territory as Turkey did in the lead up to the invasion of Iraq in 2003. This led to a branch plan of sailing an entire division through the Suez Canal to Kuwait. Elements of the plan changed without changing the overall operational and strategic objectives.

Throughout a football game or a season, things don't always work out as expected. Players get injured, other teams evolve and innovate their formations and play calling, and other times plays and game plans fail for a myriad of reasons that a coaching staff may not grasp due to the complexity of interactions on the field. Coaches and players anticipate these unexpected turns to the game ahead of time. Quarterbacks for example often call audibles or play changes at the line of scrimmage should a defense line up in specific ways or have certain players on the field. Coaches can adjust their game plans by drawing out new schemes on the sideline or in the locker room during halftime. As the season unfolds general managers may look to trade players away should it become apparent that the current roster is not competitive. Comparably, a general manager may trade for or sign a player mid-season if they think they are one piece away from a championship run. Branches can help to prevent complete failure and to reinforce success.

Important to understanding branch plans is that commanders, coaches, and quarterbacks must understand the capabilities of their own formations and teams. The audible, although seemingly simple, is more than adjusting the play call based on the defensive alignment. A quarterback must understand the players from his team who are on the field and their capacity to execute the new play. Coaches must have this confidence and trust in their quarterback to recognize this. Likewise, if a commander seeks to alter his or her plan, he or she must have the trust and confidence that they have the capabilities to do so and that their subordinate commanders can execute the branch plan.

Anticipation: This element of operational design is paramount for commanders who want to avoid surprise. Indeed, gaining and maintaining the initiative and forcing an adversary to react to one's actions is a competitive advantage in war.

Football coaches and players are constantly trying to anticipate what the opposing team will do next. Coaches employ scouts to observe future opponents hoping to notice tendencies or tells that might help them design their game plan. During a game, players seek out various tells based on the formations and observed tendencies of the opposing team. Defenses try to time their attack based on an anticipated snap count, while offenses try to anticipate coverage and blitz schemes of the defense.

Forces and Functions: These elements of operational design seek to have a commander design an operation or campaign focused on defeating either adversary forces, functions, or a combination of both. For example, a commander may seek to destroy the fielded forces of an enemy, much like the way U.S. Grant gradually attrited Lee's army in the Civil War. Attacking an adversary function on the other hand might lead to a focus on disruption and destruction of an enemy's ability to command and control their forces, to destroy their sustainment capabilities, or to prevent the enemy from employing long-range fires.

Aside from the elements of operational design, planners identify ***decision points*** for the commander, and for their next higher command. Understanding the decisions, a commander must make during an operation allows the staff to focus on collecting the information required for the commander to make a better and more informed decision. More than simply identifying the decisions, the best war planners think through the conditions that will lead to the decision.

Head coaches continually make decisions throughout a football game. Decisions on what to do on third and fourth down, to kick a field goal, or to go for a touchdown. To blitz the quarterback or to drop back into a prevent defense, and when to throw a challenge flag to dispute a referee's call. If the coaching staff does their job right, they have identified the possible scenarios for each decision, reducing the uncertainty and the real-time deliberation for each call. Indeed, it is frequent practice for coaches to have pre-planned the situations of when they would attempt a 2-point conversion following a touchdown, when to attempt an onside kick, and when to attempt a trick play.

The careful use of an **operational or strategic reserve** is another element not present in design, but one that must be present when commanders and staffs develop their respective plans. Maintaining a force in reserve is critical for military commanders to maintain freedom of maneuver.[3] The employment of a reserve force is paramount to either prevent failure or reinforce success. If planners and commanders could predict with absolute certainty where and when the decisive action of a battle and campaign would occur, all forces and effects would be employed in that time and space.[4] However, war is unpredictable and filled with the Clausewitzian elements of fog, friction, and chance. A reserve capability provides commanders with a tool for greater flexibility in the drive to victory.

The reserve force in football is a bit different than that in war, but no less important in the attainment of success. Backups in the second and third string can prove to be the decisive factor in a football game. Indeed, the careful replacement of key players can change the momentum of a game. More than replacing the quarterback when the starter throws a couple of interceptions, backups can provide a role when the offense or defense employs a different formation to adjust to the circumstances of a game. A second-string defensive back may be necessary to fill a role in the nickel or dime defense. A backup running back who has better skills as a pass catcher can change the game during a two-minute drill or in the attempt at a come from behind victory when the slow pace of a running game is no longer an option. Football seasons are long and violent campaigns, and teams that can best replace injured starting players tend to enjoy more successful seasons than those that cannot.

9.2 Conclusion

The elements of operational design are useful to both the military planner and football coach. More than a simple checklist of considerations when thinking through tactics and strategy, the elements provide a framework for leaders to solve complicated and complex problems, be it an enemy military force or an opposing football team.

Notes

1. *Joint Publication 5–0: Joint Planning*. (Washington D.C.: 1 December 2020). IV-1–IV-46. For this chapter, the definitions of each element of operational design are found in Chapter IV of JP 5–0.
2. Lewis Carroll, *Alice's Adventures in Wonderland and Through the Looking-Glass*. (New York, NY: Barnes and Noble Books, 2004), 73.
3. Milan Vego, *Joint Operational Warfare: Theory and Practice*. (Newport, RI: U.S. Navy War College, 2007), VII-121.
4. Ibid.

Chapter 10
Revolution, Innovation, and Adaptation

If necessity has been the mother of invention throughout the history of professional football, it could also be said that desperation is the father. – Doug Farrar.

There are two ways to fight the United States military, asymmetrically and stupid. - H.R. McMaster

A revolution in military affairs (RMA) is "the transformation of a military establishment's strategies, doctrines, training, education, organization, equipment, and operations and tactics to achieve decisive military results in fundamentally new ways."[1] Military history provides multiple examples of these RMAs. These examples include the development and use of the English Armies six-foot-long bow and the use of gunpowder artillery in the 14th Century. The French use of *levee en masse*[2] during the Napoleonic Wars, the German development of combined arms in the form of blitzkrieg during the Second World War, and the advent of nuclear weapons provide further examples.[3] More recently, the development of precision-guided munitions enabled by military capabilities in space, stealth technology, and unmanned and autonomous systems serve as modern day examples.

The concept of a revolution in military affairs easily translates to football. The past 70 years witnessed the development of new offensive and defensive systems (operations and tactics) that transformed the character of the game. These transformations began with the development and implementation of the forward pass, whose first large scale use enabled Pennsylvania's Carlisle Indian Industrial School coached by Pop Warner to defeat the University of Pennsylvania in 1907. Six years later, in 1913, the forward pass helped Notre Dame defeat the highly favored Army team.[4] Other offensive innovations throughout the years include the West Coast Offense, the Run-and-Shoot, the K-Gun Offense, and the read-option employed by the Super Bowl winning 2017 Philadelphia Eagles. Nevertheless, football has a lesson to teach concerning rapid changes that provide an advantage to one side of the field. Opponents quickly adapt to counter or limit the advantages posed by new schemes.

D. Sukman, *American Football and the American Way of War*,
https://doi.org/10.1007/978-3-031-55345-5_10

10.1 Innovation

Following the 1973 Yom Kippur War, the United States Army realized that it needed a vision and doctrine to defeat the massive formations of the Soviet Army. Senior army leaders led by General William DePuy used the lessons learned to create the doctrine of Active Defense. Shortly thereafter DePuy's successor General Donn Starry created and implemented AirLand Battle doctrine, which came to successful fruition in the 1991 Gulf War. Army doctrine has evolved and adapted since, to include Full Dimensional Operations, Full Spectrum Operations, COIN, and the current efforts for Multi or All-Domain Operations.[5] Innovation and revolutions in war don't necessarily require things or materials as much as they require thinking and brainpower.

Often it is the lack of money or material items that force nations and armies to use their collective brainpower and change the way they fight wars. Blitzkrieg, or the operational approach Germany used early in World War II came to fruition through the thinking of military leaders who were unable to develop and experiment with new military technology due to the restrictions outlined in the Treaty of Versailles. The imposed limitations placed German military leadership in a position to create new and innovative operational approaches to battle before developing the combat platforms to fulfill these ideas. Relying on operational success however can have drawbacks, as Germany completely failed at the strategic level, fighting a way with few to zero allies while fighting against adversaries who possessed vastly superior amounts of personnel and material.

Innovation isn't limited to the land domain. New technologies and ideas in the maritime domain also come from necessity. When Admiral Wylie, who we discussed in Chapter 4 was a mere Lieutenant Commander, he was central to operating the Navy's first Combat Information Center (CIC). As Trent Hone detailed in Learning War, early on in the Pacific Theater of World War II, ship commanders were "unable to quickly make sense of a developing situation and act on the available information to them…the cognitive load was too high."[6] The CIC was the tactical innovation that would prove to be key to U.S. Naval victories over Japan, and remains to this day a central feature aboard U.S. Navy warships.

Just as one nation transforms the character of its military and warfighting, other nations will respond with changes to counter said advantages. In football, rapid changes in offensive concepts drove changes to opposing defensive systems. One example is Buddy Ryan's ferocious 46 defense.[7] This was a defensive system employed by both the 1985 Chicago Bears and 2000 Baltimore Ravens, two of the greatest defenses of all time.[8]Buddy Ryan, who before his coaching career served as an Army NCO in the Korea War, conceptualized the 46 defense when he realized his team didn't have the talent to apply pressure on the quarterback using standard 3–4 and 4–3 defensive schemes.[9] The application of this defensive scheme was out of necessity and in response to offensive systems employed by the San Francisco 49ers.

A second example of defensive innovation is the zone blitz, designed to counter the Run-and-Shoot. The founding father of the Zone Blitz is Dick LeBeau who was seeking a method to apply pressure on the quarterback without forcing the defensive backs (cornerbacks and safeties) to cover the entire field.[10] Defensive counters to offensive weapons, be it football formations or anti-access area denial (A2AD) systems such as patriot missiles are a part of the higher-level strategic dialogue between adversaries.

Changing warfighting doctrine is similar to the way successful football coaches adapt their style of play to match the talent on the team. Since the early days of football, some of the central formations included the wishbone, the T, the power I, and the pro-set to name a few. In the 1990s, the Run and Shoot offense was a powerful innovation to account for the talents of Warren Moon and his fast wide receivers playing for the Houston Oilers. In the mid-2000s Bill Belichick implemented a two-tight end set to account for the unique talents of Rob Gronkowski and Aaron Hernandez.[11] More recently, teams such as the Kansas City Chiefs have been employing the Run Pass Option (RPO), which forces a defense to account for the quarterback who has the option to hand the ball off, run it himself, or throw to a receiver downfield. Indeed, players with the skills of quarterbacks such as Patrick Mahomes who can run, and throw serve to create multiple dilemmas for the defense.

Perhaps one of, if not the most influential innovations in football was the T formation. It was Clark Shaughnessy who first implemented the system at Stanford in 1940. Like many innovations and adaptations, Shaughnessy did not invent the formation but understood how to use it with the players available to him. In his book *The Wow Boys*, James W. Johnson explains how Shaughnessy leveraged the warfighting concepts employed by Heinz Gudarian on the Western Front to develop plays within the T formation.[12]Indeed, Shaughnessy compared older methods of football offense to the trench warfare of the First World War, where armies made progress foot by foot, yard by yard in a geographically constrained battlefield. With the T formation, Shaughnessy's Stanford teams forced the defense to defend the entire field, opening space for an offense to choose its point of attack.[13] Forcing an adversary to defend everything is an effective method to have them truly defend nothing.

Innovations don't last forever. Sometimes the innovation is unique to the talent and capabilities of an organization, and other times adversaries adapt and develop their own ideas to counter and overcome the original innovation. Other times, the innovation is simply not sustainable over the long term. The Run and Shoot for example assumes risk in other parts of the game, as offenses had to make room for an additional wide receiver on the roster which came at the expense of a tight end or full back, positions that often fill critical roles on special teams. More recently, Chip Kelly's attempt to implement his Blur offense failed to take root as opposing defenses caught up to his schemes. Indeed, Doug Farrar in his book *The Genius of Desperation* recounts how defenses came to understand that the quarterback had no time to audible, and thus would hide their coverages.[14] In any endeavor where there is a living and thinking adversary change is a necessary constant.

An example of a unique capability or talent that requires innovation or adaptation in football is the case of Lawrence Taylor. In the early 1980s, Taylor in his blue

#56 New York Giants uniform seemed to be an unstable force, wreaking havoc in the backfield of opposing offenses. The counter to Taylor came in the form of a two-tight end formation with one back. This allowed for the double, and sometimes using three lineman (triple teaming) to block Taylor.[15]Joe Gibbs, head coach of the Washington Redskins employed this formation often having the tight end or fullback line up behind the tackle on Taylor's side. Gibbs' teams would take this innovation to multiple Super Bowl victories.

Technology also plays an influential force in how a football team develops and implements its playbook and week-to-week game plans. An observation of the sidelines today reveals wireless communication between coaches and players on the field. Further, players have instant access to video reviews and overhead imagery as they look at tablets while sitting on the sidelines. This technology speeds up the rate of learning so that teams can adjust their plans in real time. These real time information technologies mirror developments on the battlefield, as today commanders rely on real time satellite imagery and instant communications to subordinates to inform their decision making.

The constant evolution and innovation swaying back and forth between offense and defense is similar to how the United States, and in large measure, the West thinks about military innovation. One side develops a new technology, weapon, or tactic, and the other side develops a counter technology to offset any tactical, operational, or strategic advantage. Indeed, this continual back and forth resembles chess more so than Eastern culture type games such as Wei. Teams consistently try to stay one step ahead, on the field and off.[16] Promises of technology or ideas that will change the nature of warfare and provide one nation a permanent advantage are often nothing more than a fantasy.

The U.S. military looks for advantages in more ways than technology alone. For example, improvements to training methods and physical fitness are a cornerstone of combat preparedness. The U.S. Army's recent implementation of the Army Combat Fitness Test (ACFT) for example seeks to change the culture of fitness in the army from one of aerobic endurance to one that balances endurance with muscular strength. Changes to football occur off the field too. Long gone are the days when professional athletes used training camp and the pre-season to get in shape. The advancement of fitness methods and technologies allows players to train throughout the year. Coinciding with fitness training are new understandings of proper diet combined with the proper use of supplements. The Steelers Dynasty of the 1970s totaled zero players over 300 pounds.[17] Today, 300 pounds is the minimum weight to play as a lineman in the NFL. Of course, influencing the evolution of the size of players is the rules of the league. In the early era of football, players often remained on the field for an entire game as free substitutions were not allowed. Once the rules changed to allow for free substitutions, the era of two-platoon football and position specialization came into being.[18] Players today are bigger, faster, and stronger than in any previous era, but it is doubtful a 340-pound lineman could run up and down the field for 60 minutes playing iron-man football.

Often, what leads to innovation and adaptation are changes to the rules of the game. In his book, *How Football Became Football*, Timothy P. Brown details eleven major changes to the rules that brought football into the modern era. These rule changes include a system of downs, the implementation of a line of scrimmage, the introduction of the forward pass, hash marks, the use of unlimited substitutions, television money, and racial integration. Of note, when football first began, when a player came off the field for a substitute, said player could not reenter the game.[19] When this rule changed, teams began using a two-platoon system leading to specialization of players for their positions.[20] Half-backs no longer had to play defense and quarterbacks were no longer required to punt and kick. The increased specialization led to larger coaching staffs who could focus specifically on offense, defense, and special teams, as well as the position coaches used in the modern era.

In another example of the relationship between war and football, it was the military and World War II that drove the necessity for two-platoon football. During the war, about a third of the league was called into service with 22 current and former players making the ultimate sacrifice.[21] With a dearth of skilled players who had left the gridiron for military training, college and professional teams needed more leeway to substitute for injured players, thus forcing the rules committee to approve this change in 1941.[22] Necessity is indeed the mother of invention.

As an aside, it was Red Blaik, coach of the West Point football team who coined the term "two platoon" to indicate the offense and defensive players.[23]Blaik understood the term from his time serving in the Army during the interwar years between WWI and WWII, and of course from the military structure of West Point.

Larger coaching staffs are the equivalent of the rise in the size and scope of headquarters in the military. In war, as weapons evolved to shoot further distances, and the complexity of working in additional domains including the air, space, and cyberspace domains became normal, commanders became more reliant on staffs that held expertise in each area. The staffs of commanders such as Washington and Grant were minuscule compared to headquarters of commands such as U.S. Central Command, U.S. European Command, or even the service headquarters in the Pentagon that number in the thousands. General Grant and his staff never analyzed satellite imagery of enemy positions, nor did they contend with confederate malware infiltrating command and control communication systems.

Innovation in football is not confined to the rule books and playing field. Part of the NFL's dominance as America's sport lies in its ability to recognize areas for profit. The NFL was the first professional sports league to launch its own website and run a 24-hour television network.[24] Other professional sports leagues would follow the NFL's lead, and today every major professional sport runs its own streaming services, thus providing fans the ability to watch any team at their own convenience. The ability to exploit new technology is a mark of an innovative culture.

Evolution and innovation in football and war highlight other similarities already discussed in this chapter. The back and forth between opposing sides demonstrates once again how war is fought against a living and thinking adversary, the law of reciprocity expressed by Clausewitz. Moreover, adaptation between two thinking adversaries occurs at each level of war, the tactical, the operational, and the strategic.

In warfare for example, the First World War was witness to a continual search for new technologies and warfighting tactics to break the stalemate between the

Allies and Central Powers. The Germans developed auftragstaktik, or ways of infiltrating enemy lines with small forces while bypassing enemy strong points. The Allies evolved with the development of new technology such as the tank. In the latter example, the Allies were successful at the Battle of Passchendaele. Following WWI, the Germans continued to adapt with a doctrine that massed tanks and emphasized mobility, leading to a successful campaign against France in 1940. As already mentioned, the massing of armored formations with a pointed attack for exploitation would serve as a model for football's T formation.

At the tactical level, quarterbacks audible plays while coaches make in-game adjustments to their play calling and formations. At the operational level, coaching staffs change their game plans week to week to account for opposing teams' strengths and weaknesses. At the strategic level, teams draft, and trade for players to build their strength, and to counter the strengths and weaknesses of opponents in their divisions and conferences. Innovation and adaptation are continual processes in both war and football.

The 19th, 20th, and twenty-first centuries saw innovation and adaptation at the operational level of war. Indeed, it was in the 19th Century when the development of the operational level of war began to influence military leaders and planners and was itself an innovation in thinking. General Grant in the U.S. Civil War came to understand that the defeat of the traitorous confederate army would not occur in a single decisive battle. Rather, Grant was able to link a series of battles across space and time, leading to the surrender of Robert E. Lee and his army at Appomattox. In the 20th Century, new domains such as air, space, and cyberspace became considerations for military leaders in their planning. Joint and combined arms warfare is not a luxury, but rather a necessity for operational level success.

The 20th Century was also witness to multiple strategic level revolutions. The advent and subsequent proliferation of nuclear weapons was the first, and the transition to precision and stealth warfare was the second. The introduction of nuclear weapons on the battlefield at the end of the Second World War ushered in a new era of warfare, one where the great powers understood that conventional military conflict could escalate to a nuclear confrontation. Precision guided munitions and stealth airplanes led to a reduction of civilian casualties, but also to an expectation of zero defects and zero civilian casualties. Moreover, the ease and distance from which modern day combat platforms can reign lethal effects upon an enemy can make the decision to use the military to solve problems all the easier. Indeed, the military arm of national power often becomes the hammer in an environment where all problems then resemble a nail.

The institutional level of war is perhaps the level where the most profound changes occur for a military force. Building a military is more daunting and complex than fighting a military, and is filled with more uncertainty, as institutional planners must predict the future from anywhere from five to thirty or more years out. Warfighting platforms in each of the services, from B-52 bombers to M1 Abrams tanks can have lifespans that last decades with service in multiple wars in various places across the globe.

The ability to predict future performance is essential to building a successful franchise and winner on the field. How to properly evaluate talent and predict future performance is one of the lingering questions of both team sports and the U.S. Military. In football, the advent of big data and advanced statistics allows coaches to better judge performance on the field. Indeed, in the modern era, every player chosen in the annual draft has been scouted and evaluated by coaching staffs. These evaluations range from observing in game performance, interviews, and often a series of physical evaluations at the NFL combines. Taking lessons from the NFL, the U.S. Army in 2019 held its first ever Battalion Commander Assessment Program (BCAP). Modeled after the NFL combines, BCAP evaluated potential battalion commanders in a variety of events from an interview to a writing exam to a physical fitness test. This evaluation concept expanded to include a Colonels Command Assessment Program (CCAP) for those seeking brigade level command.

So who is more responsible for success; the head coach who devises offensive and defensive schemes employed on game day, or the general manager who drafts and trades for the talent the coach employs? Is it the general who built the army or the general who draws the arrows on a map to fight the army? The answer of course is that the best football organizations and the best militaries have institutional and operational level leaders who work together towards common objectives balancing the acts of building and employing teams and military organizations.

10.2 Innovation Failures

New technology alone is not enough for successful innovation. New technologies and combat platforms must come with innovative ideas for how to organize and employ them in combat. One of the more notorious examples of failing to develop appropriate doctrine with new systems was the French in the lead up to World War II. While the Germans were testing tanks in new armored formations that could rapidly maneuver on the offensive, the French dispersed their tanks to support infantry centric, static defensive positions on the French-German border. When the Germans finally attacked in 1940, it took all of six weeks to get to Paris. The French had better tanks, but the Germans had the better imagination on how to use them.

Sometimes a military or a football team will develop new and exciting concepts, only to have them fail spectacularly in practice. The key to successful innovations in play design is to have the right personnel and talent to execute the new ideas. The U.S. military's recent experience with concepts such as Effects Based Operations, commonly referred to as EBO. Effects Based Operations was the U.S. military's attempt to apply systems thinking to the battlefield. This idea, however, quickly devolved into an attempt to turn the complications of war into a simplistic targeting methodology. War, however, is not a targeting game and what looks good on the chalkboard can often translate into disaster in practice.

When former Secretary of Defense James Mattis commanded U.S. Joint Forces Command as a 4-Star General, one of his first instincts was to rid the joint force of

Effects Based Operations. According to Mattis, EBO attempted to place certainty on the uncertain nature of war. It was a method that worked well for the U.S. Air Force but could not work for all the services.[25] According to Mattis, EBO centralized command and control, and turned soldiers and marines into "reporters on the battlefield," leading them away from the warrior mindset. Further, EBO required predictions or forecasts on the second and third order effects of each military action. In an activity defined by uncertainty and human decision making this is a near-impossible undertaking.[26] While the character of war evolves, the nature of warfare remains constant. EBO and other attempts to reduce and eliminate the fog, friction, and chance inherent to war continue to be fruitless (Fig. 10.1).

Changes to doctrine and organization take time. New doctrine should begin as a concept or idea and undergo rigorous testing and evaluation before it is implemented to the force. The U.S. Army witnessed failures in new organizational and warfighting doctrine efforts. For example, the creation of the Pentomic divisions in the 1950s. Here, infantry and airborne units were organized with the idea of five subordinate units for each level of command. The hope was that this organization could survive enemy use of tactical nuclear weapons on the battlefield. The flaws of this organization were the increased span of control combined with the elimination of the battalion level of command. The pentomic division did not last long, ending less than ten years after its implementation.

Despite the failures of the Pentomic Division and Effects Based Operations, the U.S. Army has been surprisingly successful at doctrinal and organizational change

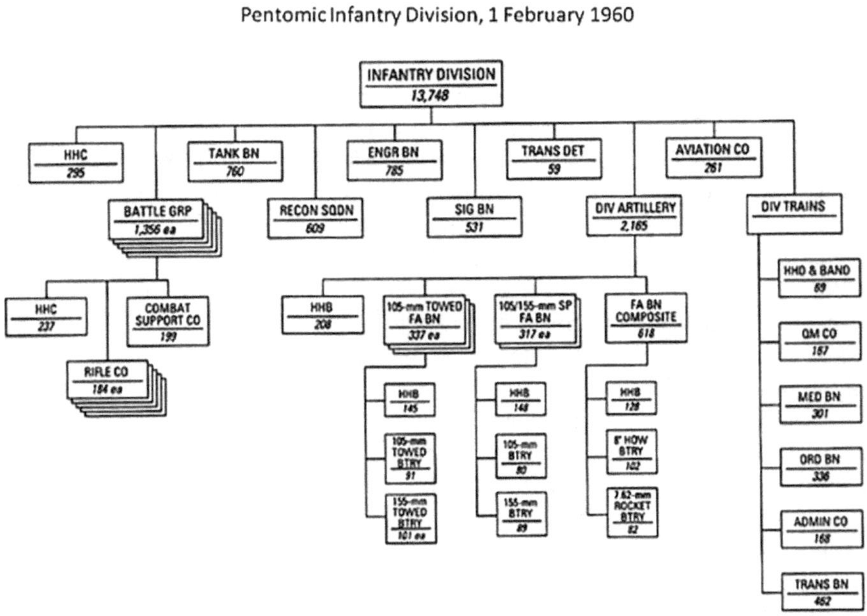

Fig. 10.1 The Pentomic division (*Source* United States Army Center for Military History)

over the last forty or so years. As mentioned previously in this chapter, Benjamin Jensen, in his book *Forging the Sword* captured how the Army, from the early 1970s through the second decade of the 21st Century revised its central operating doctrine seven times. These doctrinal changes included Active Defense, AirLand Battle, Full Dimension Operations, Full Spectrum Operations, Counterinsurgency, and Multi-Domain Operations.[27]

Failures of innovation are not the sole domain of warfighters. There are many examples of failed innovations in football from offensive and defensive schemes to experiments with changes to the rules. In the latter, in 1989, the NFL inserted a rule to regulate the noise of the home crowd by penalizing the home team when the visiting team's quarterback could not hear or communicate signals to the rest of their team. Aside from the ridiculous thought that the league should penalize fans paying to see a game for cheering too loud, the use of wireless communication between the coach and quarterback combined with silent counts rendered the rule mute.

At times it is not the ideas that fail, rather it is leadership that does not have the foresight to understand when change is looming on the horizon. Other times there is an attachment to a flawed idea. Leaders who are set in their ways and comfortable with the status quo often serve as barriers to organizations that need to change. During the interwar years, General John K. Herr, the chief of cavalry, failed to see the utility of motorized vehicles and tanks. Indeed, General Herr insisted that horse cavalry would remain relevant despite the lessons of the First World War and mechanization of European armies, even publishing and advocating for the need for more horse cavalry in professional journals as late as 1939.[28] While there were officers in the U.S. Army such as Dwight Eisenhower and George Patton who understood the changing character of war, their voices were squashed, to the extent that Eisenhower stopped publishing his ideas under threat of a court martial.

Other inhibitors of innovation include a fear of failure. When the leadership of an organization, be it a military unit or a football team, preaches a return to the fundamentals as the way forward, new ideas simply fail to materialize. The organization becomes set in its ways, seeking to perfect tried and true methods. In an activity where an adversary has observed these methods and is thinking how to defeat them, remaining, or wishing to return to a "golden age" is a recipe for long-term failure. Leaders of military and football organizations alike must be champions of ideas and innovation and are the keystone to preventing stagnation.

Military organizations and football teams can overcome barriers to innovation through several methods. When diversity in background, education, and thought is prevalent new and innovative ideas tend to spread. In his phenomenal book *Military Adaptation In War: With Fear of Change*, military historian and prolific author Williamson Murray describes how organizations require imagination and willingness to change as key cultural factors for organizational success. Moreover, he details how successful organizations solicit input up and down the hierarchy.[29]

Balancing hierarchy and the chain of command with a culture where anyone can propose ideas and solutions to problems is an effective means. As an example of the latter, prior to Super Bowl LI Brian Hoyer, the Patriots' back-up quarterback was able to pick up on a tell of the Los Angeles Rams. His insights were communicated to the coaching staff who used it in their game plan.[30] As a result, the high-scoring Rams were limited to a single field goal and never made it past the Patriots 20-yard line. Within innovative organizations, nobody has a monopoly on good ideas.

Leadership empowering a diverse team is one aspect of many within successful innovation. In his book *Leading Change*, author John P. Kotter wrote extensively on the importance of leadership. Leaders create and continually communicate the vision, leaders create and guide an active coalition for change, and leaders create a sense of urgency in the organization on the necessity of change.[31] Leaders are ultimately responsible for what their organizations accomplish, and what they fail to do. The best leaders create the environment for innovation and change ensuring sustained and long-term success.

10.3 Conclusion

Successful military organizations were able to think through innovations in peacetime, where resources and funding tend to be scarce, and then adapt during war when the firehose of money turns on.[32] There is an ongoing dialog between tactical, operational, and strategic leaders who can visualize change, communicate the vision for change, and implement necessary change. However, any advantage gained through adaptation and innovation is fleeting, as enemies and adversaries are continually learning and adapting at the same time.

The decisive advantage in both war and football is the ability to learn faster than your adversary. New tactics, techniques, and procedures (TTPs), and advancements in technology influence the character of both war and football. Militaries and teams that can devise new and innovative ways to take advantage of their inherent strengths and overcome their weaknesses will likely prevail on the battlefield and on the gridiron.

Notes

1. Michael Keane, *Dictionary of Modern Strategy and Tactics*. (Naval Institute Press. Annapolis, MD 2005).
2. Levee en masse is the practice of universal conscription. First established in 1793, this method enabled France to fight 14 armies, an unheard-of number at the time.
3. Michael Keane, *Dictionary of Modern Strategy and Tactics*. (Naval Institute Press. Annapolis, MD 2005).
4. Jim Morrison. 2010. "The Early History of Football's Forward Pass." *Smithsonian Magazine*. Accessed 23 October 2018. (https://www.smithsonianmag.com/history/the-early-history-of-footballs-forward-pass78015237/).
5. Benjamin M. Jensen. *Forging the Sword: Doctrinal Change in the U.S. Army*, (Stanford, CA: Stanford University Press,2016), 4.
6. Trent Hone, *Learning War: The Evolution of Fighting Doctrine in the U.S. Navy, 1898–1945*. (Annapolis, MD: Naval Institute Press, 2018), 206–208.
7. The 46 defense was named for the jersey number of Chicago Bears safety Doug Plank and has nothing to do with the number of linemen, linebackers, or defensive backs on the field.

8. Although statistically the 1986 Bears' defense was better, the team lost to the Giants in the playoffs and did not win the Super Bowl. Thus, the 1985 Bears became the common reference point of great defenses.
9. Tim Layden, *Blood, Sweat, & Chalk: How the Geniuses of Football Created America's Favorite Game.* (New York: Sports Illustrated Books, 2010).
10. Ibid.
11. Aaron Hernandez was a unique talent that helped the Patriots advance to the 2011 Super Bowl. This is a strict judgement on his football talent and not his tendency to murder people in the off season.
12. According to the author, Shaughnessy would not admit to this until the 1960s, when he finally considered it safe to admit a nazi general's tactics was his inspiration.
13. James W. Johnson, *The Wow Boys: A Coach, A Team, and a Turning Point in College Football.* (Lincoln, NE: University of Nebraska Press, 2006).
14. Doug Farrar, *The Genius of Desperation. The Schematic Innovations That Made the Modern NFL.* (Chicago: Triumph Books, 2018). Kindle.
15. Matthew Sherry, *Any Given Sunday: The NFL's Epic 100-Year History in 20 Games.* (London, UK: Orion, 2020). Kindle, Loc 4218.
16. Ron Jaworski, *The Games that Changed the Game.* (Random House Inc. New York, 2010).
17. According to Pro Football Reference, the heaviest player on the 1976 Steelers was Joe Greene, who stood 6′4″ and weighed 275 pounds. Compare to the 2018 New England Patriots Left Tackle, Trent Brown who stand 6″8″ and weighs 380 pounds. (https://www.pro-football-reference.com/).
18. Timothy P. Brown, *How Football Became Football: 150 Years of the Game's Evolution.* (West Bloomfield, MI: Brown House Publishing, 2020).
19. Ibid.
20. Ibid.
21. Sherry, Loc 986.
22. Timothy P. Brown, *How Football Became Football: 150 Years of the Game's Evolution.* (West Bloomfield, MI: Brown House Publishing, 2020).
23. Ibid.
24. Sherry, Loc 5922.
25. Jim Mattis and Bing West. C*all Sign Chaos: Learning to Lead* (New York: Random House, 2019). Pg 180–183.
26. Ibid.
27. Jensen, 4.
28. David E. Johnson, *Fast Tanks and Heavy Bombers: Innovation in the U.S. Army, 1917–1945.* (Ithaca, NY: Cornell University Press, 2013), 138.
29. Williamson Murray, *Military Adaptation in War: With Fear of Change.* (Cambridge: Cambridge University Press, 2011). Kindle.
30. Mike Cole, "How Brian Hoyer Played a Pivotal Role in Patriots Defense's Super Bowl Success. 11 February 2019. Retrieved from: https://nesn.com/2019/02/how-brian-hoyer-played-pivotal-role-in-patriots-defenses-super-bowl-success/
31. John P. Kotter, *Leading Change.* (Boston: Harvard Business School Press, 2012), 21–29.
32. Williamson Murray, *Military Adaptation in War: With Fear of Change.* (Cambridge: Cambridge University Press, 2011). Kindle.

Chapter 11
Civil-Military Relations

> the general must understand that he is not a free agent and cannot hope to become one. He has to work within the limitations imposed by the fact that he is working for a democracy, which means that at times he must modify or abandon the soundest military plan and make do with a second-best.
>
> – Bruce Catton, Mr. Lincoln's Army.
>
> Look, I don't want to win in the press. I want to win on the field. Al Davis

One of the principal considerations at the strategic level of war is the interaction between civilian leadership and military officers. This relationship is fundamental to how a nation prepares for and fights its wars. There are two predominant theories of civil-military relations. The *normal theory* espoused by Samuel Huntington, and the *unequal dialogue* proposed by Eliot Cohen.

The normal theory highlighted in Huntington's *The Soldier and the State* postulates that civilian leadership is the group that is responsible for when and where a nation will use the military instrument of national power to achieve or maintain its respective national interests. Further, this normal theory espoused the idea of objective control, where the nation isolates its service members from politics, thus giving them absolute freedom in deciding how to fight a war.[1] In essence, civilian leaders remain strictly at the grand-strategic level deciding when and where the military will fight, but leaving the operational and tactical levels to the military. However, this idea runs in direct contradiction to Clausewitz's theory that war is an extension of politics, and thus politics and war are two sides of the same coin.[2]

It was Eliot Cohen who popularized the concept of the unequal dialogue, which maintains that civilian leaders have the absolute right and mandate to continually make strategic level decisions, including those that directly impact the military. This includes placing limitations on the extent of military force (rules of engagement), the hiring and firing of military commanders, decisions on the number of forces, and the priority of military efforts.[3] The idea of the unequal dialogue is paramount in the modern nuclear and information age where the way in which a military fights a war can sway domestic and foreign opinion of the war. There are two famous examples of this unequal dialogue in American military history.

D. Sukman, *American Football and the American Way of War*,
https://doi.org/10.1007/978-3-031-55345-5_11

Throughout the American Civil War, President Lincoln continually involved himself in military decisions. For example, early in the war, Lincoln hired and fired military commanders such as George McClellan and Joseph Hooker for poor performance until he was able to find a commander who understood his intent to destroy the Confederate will to fight through the destruction of Confederate armies. This commander was Ulysses S. Grant. In the Second World War, President Franklin Roosevelt was the final approval authority for who would lead the invasion of Europe and command Operation Overlord. It was Roosevelt who informed General George Marshall that he was too valuable to the strategic direction of the war, centered in Washington D.C. to go overseas.[4] Roosevelt decided to appoint Eisenhower as the commander of Operation Overlord. Further, it was President Roosevelt who overruled Marshall's desire for an early invasion of Europe in 1942, favoring operations in North Africa and the Mediterranean. In each of these cases, the President overruled his subordinate generals with the understanding that the conduct of the war was just as important as the decision to conduct the war.[5] In arguably the two wars fought for America's survival as a nation, civilian oversight was critical to victory.

Throughout America's history, there are numerous examples breakdowns in the civil-military relationship. Perhaps the most famous example is President Truman's firing of General Douglas MacArthur at the height of the Korean War. In this example, General MacArthur repeatedly made public statements in opposition to President Truman's policy vis-à-vis the Korean War. Indeed, MacArthur openly advocated for an expansion of the conflict and open engagement with the Chinese Military, including the possible use of nuclear weapons.

MacArthur was not the last general or admiral that a U.S. President would fire. More recent examples include George H.W. Bush's relief of the Air Force Chief of Staff, General Dugan at the onset of the 1991 Gulf War.[6] General Dugan was the focus of then Secretary of Defense Dick Cheney's ire but would not be the only 4-Star fired during his tenure. Defense Secretary Cheney fired General Fred Woerner, the commander of U.S. Southern Command in the lead-up to Operation JUST CAUSE.[7] President Obama caused a national stir with public relief of General Stanley McCrystal in 2011.[8] In the latter example, General McCrystal's staff was caught on the record making disparaging remarks towards the Vice President and other members of the Obama administration. In the American military, there is zero maneuver space for a uniformed officer to make these types of statements, or to allow his or her staff to do so.

Where civil-military relations become dicey is when senior military officials publicly lobby for their desired courses of action. For example, General David Petraeus wrote and published editorials in support of his counterinsurgency strategy and accompanying troop surge in Iraq.[9]General McCrystal lobbied for a specific increase in troops for Afghanistan through a series of public speeches and interviews, most famously on the hit TV show *60 Minutes*. What these scenarios show is that the interplay of the military and society has evolved to more than just the communications between the executive branch and the services, but also includes the legislative branch and the fourth estate.

Today, it is widely accepted that elected leaders set the nation's war aims, while military leaders devise and execute the strategy to achieve said aims. Further, political leaders provide the limitations or conditions that the military must work under. These can include limits to the number of soldiers available for an operation, governance of the rules of engagement, and determining what can be targeted with lethal or destructive force. Should military leaders disagree with their constraints, they should offer informed opinions in a respectful manner and in the proper forums.[10] Meetings with executive branch officials and testimony to Congress are two examples.

In football, there are two similarities to civil-military relations. This is the relationship between the head coach and the quarterback, and the relationship between the owners, general managers, and head coaches. Just as the President can fire military commanders in their role as Commander in Chief, owners of professional sports teams exercise similar authority. In recent years, some owners tend to act more aggressively in the hiring and firing of head coaches. Dan Snyder of the Washington Commanders and Jimmy Haslam of the Cleveland Browns serve as examples of owners who regularly changed the leadership within their organizations. Other owners such as the Rooney family in Pittsburgh and Bob Kraft in New England value continuity in the franchise and take a more hands-off approach. Not surprisingly, teams that play in a stable organization tend to have more long-term success.

In professional football, the team owner has ultimate control of the team. This control includes retaining decision authority for hiring and firing of the coaching staff, awarding contracts to the players, and deciding on who the team will pick in the annual draft. Often, the head coach advises the owner and general manager on who to draft, and what players to sign in free agency. However, the owner is free to heed or disregard the general manager and head coach's advice. Thus, the relationship between an owner and a head coach is an unequal dialogue. A head coach has the responsibility to offer candid advice and to direct the action on the field as he sees fit, but the owner has the final authority in every decision, from hiring and firing coaches, players, and support staff to calling the plays on the field. An owner has the absolute authority to hire and fire the head coach, and any subordinate member of the coaching staff, training staff, and the signing and releasing of players.

On the field, it is generally accepted that the head coach, along with the offensive and defensive coordinators, is the final authority for game decisions. These decisions include play calling, whether to kick or punt on fourth down, and when to employ a trick play. While some coaches may place a greater amount of trust in their players to adjust or audible the play call, like the owner-head coach relationship, the coach has the absolute authority to direct the action on the field during a game.

On occasion, the player-head coach relationship breaks down. During these events, the outcome is often dependent on the stature of the player. The star quarterback, wide receiver, or linebacker tends to have more leeway than the backup offensive guard who rarely steps on the field during a game. In some cases, the breakdown can lead to the release of a player from a team, while in other cases the owner may be more invested in a star-player, and choose to fire the head coach. Breakdowns in interpersonal relations are not unique to the gridiron and lead to the idea of principled resignation where a subordinate should resign from their position if they cannot carry out the directed policies of their leadership.

Often, the idea that senior leaders who make decisions on policy, and when and where to use the military instrument of national power should have military experience and have served in uniform comes into the light. However, the simple counter to this is that two of the greatest wartime national leaders, Abraham Lincoln and Franklin D. Roosevelt had never served in the active military. Despite never having served, both would continually overrule their generals and admirals and on many occasions relieve leaders they found to be incompetent. As the historian Williamson Murray noted, "time spent in the service is not a prerequisite to understand the complexities of war, or in the making of strategy."[11]

Similarly, spending time playing football is not a prerequisite to understanding the complexities of game planning, nor is it necessary to understand the business aspects of the game. Bill Belichick for example was a special teams player at a low-level college before he broke into the coaching profession. His team owner and boss never played at the competitive level, yet both were able to combine for six championships during their time together running the New England Patriots franchise. As David Halberstam noted in his biography of Belichick, in football, what matters most is knowledge. It's more important than the school you played at, and more important for a coach than a loud voice.[12] Talent and intelligence are greater than pedigree.

The idea that senior leaders should resign is a complicated subject matter in the military. Dwight Eisenhower, while serving as Supreme Allied Commander in Europe threatened resignation on multiple occasions, most notably in the days leading up to the Normandy invasion as the British were reluctant to support the Transportation Plan, Ike's method of destroying roads and railways to prevent German reinforcement of positions in France.[13] General George Marshall and his British counterpart Sir Alan Brooke both threatened to resign should the planned invasion of Sicily have been called off.[14] More recently, in 1997 General Ronald Fogleman, the Air Force Chief of Staff resigned from his position based on policy disagreements with then Defense Secretary William Cohen. Ultimately, if military leaders cannot support policy to the best of their ability, or if they feel that their advice is not sufficient, and another officer is better positioned to provide military advice, they should resign.

There are times in football organizations when coaches and/or general managers should understand that it is time to resign and seek employment elsewhere. When owners hire general managers who have a different philosophy on how to build a successful football team that is diametric to the head coach's system, then it is time for the head coach to find a new employer. Often, coaches are given a specific timeframe to implement their plan and to build a team into a consistent winner, and into a champion. Failing to meet the desired timelines of an owner, or to show continued and significant improvement often serve as a criterion, or as an indicator that an owner will seek to change the leadership within the organization.

A critical aspect of civil-military relations is one of trust. Elected and appointed civilian officials must trust that their uniformed counterparts will always offer their best and informed advice on the use of the military instrument of national power.

Conversely, the military trusts that civilian leaders will not politicize the ranks to advance other policies or political agendas. Moreover, the military trusts that their elected officials will provide policy guidance, or war aims when employing the military. Unfortunately, not every leader is a Lincoln, and there are times when policy guidance is unclear or even contradictory.

A final consideration in civil-military relations is the appointment of former generals and admirals to senior-level civilian positions in the United States Government. While America has a rich history of former generals serving in office, from George Washington to Dwight Eisenhower, the recent appointments of James Mattis and Lloyd Austin to serve as Secretaries of Defense in the Trump and Biden administrations opened some fissures. Each required a waiver from Congress to the 1947 National Security Act, which prohibited retired military officers from serving in that position for a period of 7-years after retirement.

Sports, and football, often look to fill the coaching ranks with former players. This thinking goes beyond the sideline, as television networks often hire former players as commentators for their programming. However, not all the great coaches played at the highest levels of the game. Further, only looking at former players eliminates approximately fifty percent of the population in the search for talent.[15] Both football and the military experienced their own struggles to build an inclusive environment concerning gender and race.

Effective civil-military relations require transparency not only with public officials but with the general population. Although the military should not share war plans and other sensitive material with the public, open dialogue on recruiting and retention policies, or acquisition programs helps establish and maintain trust. Further, this dialog must be honest and forthright, and generally avoid situations such as the Five O'Clock Follies of the Vietnam era. Viewing the press as an adversary, enemy, or tool to shape and influence domestic public opinion rapidly erodes trust in those in uniform. This trust or lack thereof has a direct impact on the ability to recruit and retain service members and the ability to meet mandated end strength.[16] Armies don't win wars, nations do. And a population that cannot fill the ranks of its military is a population of a nation that will lose its wars.

Like the military, league officials, coaches, and players should seek to form a positive relationship with the public. At Pop Warner and High School levels, this would include dialog between coaches, players, and parents. All of whom influence and decide on the participation of kids. This aspect has come to the forefront recently with the rising awareness of concussions, and the long-term damage they cause. Indeed, when parents of America's youth turn away from football as an activity, there is less talent to pull from for teams at the college and professional levels. The sport's sustained financial viability rests upon cohorts of kids who grow up playing and loving the game.

At the higher college and professional ranks, football players rely on the media to establish their brand. An adversarial relationship can damage a player's lifetime earnings in the league. Further, coaches, general managers, and owners seek to establish rapport with the fourth estate to project a positive image in the local communities. Support from fans and the local population is often key, not only to sell tickets and merchandise, and to raise television ratings, but also as a method to gain positive legislation that affects the team. Indeed, professional teams often require tax-funded stadiums, public transportation to the stadiums, and tax incentives to maintain the franchise in the community. When these relations go sour, teams can look to new cities to operate in. The history of teams such as the Raiders, Colts, Browns, Cardinals, and Rams serve as prime examples of teams packing their bags and moving to a new city.

Community relations count in the military just as does for football organizations. While football teams require a relationship with the community to build their brand and earn a profit, the military requires an engaged and informed citizenry. It is from the population that the military draws its ranks, and when Americans lose trust in military officials there is an impact on recruiting reducing the quality and quantity of men and women in uniform. Further, an engaged citizenry is more likely to hold elected officials to account for their decisions of when and where to use the military instrument of national power.

A looming impact of relations between a community and a football team or the military hinges on the economic aspects. Often there is an interdependency between a military base and the local population. Military bases bring jobs to the local area, as well as thousands of men and women who work on a federal salary looking to spend their money on everything from eating at local restaurants and drinking at local pubs to purchasing electronics at pawn shops to purchasing or renting property. When bases close the impact on the local economy can be devastating.

Football teams also have an impact on the economics of the local community. Game day at the college and professional level means fans eating at restaurants and bars or paying for parking or use of the local public transportation system. Further, special events such as playoff games and the Super Bowl at the NFL level, or the College Championship can generate significant revenue for the host city in the form of sales and hotel taxes, not to mention the part-time employment of parking attendants and food vendors in and around the stadium. Even at the high school and Pop Warner level, football games offer an opportunity for local teams and organizations to raise funds through concessions and the familiar 50/50 raffle contests. Local communities love their local teams but love them more when they win.

In the United States, the military consistently ranks as one of the most trusted professions in the country. There is never a shortage of chain restaurants or automobile dealerships offering military discounts to veterans and those still in uniform. As part of community outreach, the National Football League sponsors a Salute to Service Month each year where coaches and players roam the sidelines in military-themed hoodies and other fashion paraphernalia. Countless games including the Super Bowl kick off with a flyover of Air Force or Navy Jets and on occasion army helicopters. The use of the military to display patriotism and enhance the image of the National Football League is an image to behold (Fig. 11.1).

Elected officials and military leadership, like owners, coaches, and players all come to the table with their own experiences and expertise. Often these areas

Fig. 11.1 Navy Blue Angels flying over Super Bowl L. Photo from Defense.gov

overlap and there is friction over whose voice should reign supreme. A shared understanding of who has responsibility for the nation or the football organization leads to effectiveness, both on the battlefield and on the gridiron.

11.1 A Reflection of Society

The United States military generally adheres to the adage that the services reflect society. That is to say that each service is a diverse force composed of men and women of various religions, sexualities, ethnicities, and races, and who possess a wide range of political opinions. Service members come from all classes from the wealthy and connected to the poor and disenfranchised. However, this was not always the case as throughout history reflecting society meant excluding certain groups.

The diverse force of today is also a reflection of the modern-day accepted views of society. It is not ancient history where black service members were forced to serve in segregated units, a direct reflection of an American nation where African Americans drank from separate water fountains and attended separate schools. In her book *Homefront: A Military City and the American 20th Century*, author Catharine Lutz describes how in the 1950s, Fort Bragg, home to the Army's elite Airborne and Special Operations communities would blacklist local business for price-gauging, but not for refusing to serve black soldiers.[17] This seems like a time from a distant past, and although it's not in many of our lifetimes, it's in our parents' lifetime.

In much the same way that America brought football overseas during World Wars I and II, America also imported its racism and Jim Crow laws. In her book *American Uprising*, Kate Werran describes the friction between the British population and American service members who enforced segregation on British territory.[18] At the same time the NFL and college football were limited to white players, the American military restricted African Americans to segregated units often led by senior white officers. Up through World War II Benjamin O. Davis was the first and only African-American general officer, having been promoted to the rank of Brigadier General in 1940.[19]

Ben Davis Sr. commissioned in 1901 rising to the officer ranks from his time as an enlisted soldier. The U.S. Army wouldn't commission another black officer until his son Benjamin O. Davis Jr. graduated from West Point in 1936. Benjamin O. Davis was the first Black American promoted to general. His son, Benjamin O. Davis Jr. was the second. Before his promotion to general, Ben Jr. would command the famous "Red Tails" as a member of the Tuskegee Airman in World War II. Ben Jr. would retire as a 3-Star general while his fellow Tuskegee Airman Daniel James Jr. would become the first Black 4-Star general in 1975. In a delayed but much-deserved ceremony, Benjamin O. Davis Jr. would pin on his fourth star at a White House ceremony in 1998 (Fig. 11.2).[20]

Race problems in America and the military continued long after the conclusion of World War II. In her book *An Army Afire,* author Beth Bailey describes how the United States Army changed a multitude of regulations and policies to better integrate the service to solve simmering racial tensions that escalated on both domestic and overseas military bases. While Truman formally directed the services to integrate through an executive order, the Army struggled through years of racial strife during the Vietnam Era. This racial strife extended from the Cold War frontlines in Europe to the battlefields in Vietnam. Changes the Army instituted ranged from teaching post barbers to cut and style black hair to expanding the music selection at post exchanges.[21]

Moreover, it was not until December of 2010 when President Obama in conjunction with Congress repealed the Don't Ask Don't Tell (DADT) policy so that gays and lesbians could serve openly without the threat of other than honorable discharges, a step up from the court marshals earlier generations faced.[22] The natural criticisms of DADT were that it forced men and women in uniform to hide or lie about their sexuality. Further, throughout DADT, the military continued to discharge quality men and women from the ranks. In the book Managing Sex in the U.S. Military, Amanda Boczar points to the shocking stat that when the Army accepted moral waivers to fill the ranks during the Iraq and Afghanistan wars, 733 ex-convicts enlisted while 742 service members were discharged under DADT.[23]

More than race, color, and sexuality, the military has long struggled with acceptance of women into the force, and in combat roles. Again, it is recent changes to law that women are now allowed to serve in combat jobs, and on the front lines in combat units and positions. It was not until 2015 that Captain Kristen Griest and Captain Shaye Haver became the first women to graduate the U.S. Army's ranger school. Women are now serving in combat arms in the army, serving aboard submarines in

Fig. 11.2 Brigadier General Benjamin O Davis Sr. (*Source* Unknown photographer, Image available in the public domain)

the navy, and for a long time have been piloting fighter jets and bombers in the air force.

Whether it is race, gender, or sexuality, the inclusiveness of the U.S. military is a reminder that it is elected officials who make policy, while those in uniform execute it. While some like to extol how the military desegregated and integrated black service members before the rest of American society, it was not the military's call. President Harry Truman issued Executive Order Number 9981 in 1948.[24] We celebrate this executive order today, but must always remember that the services, and the army in particular, vehemently fought against it.[25] Indeed, General of the Army Omar Bradley, a hero of the Second World War responded by stating "the army is no place for social experiments."[26] The strategic leadership of the greatest generation that overcame the Nazis and the Japanese empire couldn't overcome their bias for upholding the existing social structure.

In many ways, sports, and in particular football serve as a mirror of American society at large. Race relations have been a contentious aspect of professional football for a long time, with the owners of the original five teams informally banning black athletes. The NFL was completely white from 1934 to 1945.[27] Despite the Redskins not integrating black athletes until 1961. Sadly, it was the desire for profits and not agreement with integration that led to change on the gridiron, as when the NFL signed contracts with the major television networks, the realization that the league could not showcase a segregated team set in.[28] Moreover, with the Redskins stadium sitting on federal land, President Kennedy threatened to refuse a lease at the DC stadium.[29] Gradually this aspect has changed, moving from 12% minority players in 1959, to 28% in 1968, to nearly 70% of the league today.[30] But despite these numbers, racial tension still exists in the league.

A major influence on the integration of black athletes into professional football was the NFL's rival league the American Football League (AFL). Before the two leagues merged in 1970, the AFL scouted players in lesser known Historically Black Colleges and Universities (HBCU) to find talent. While the NFL looked towards larger, mostly white schools to draft players, the AFL out of necessity turned to black athletes. Indeed, according to Ian S. Kahnanowich, author of *NFL 1970*, it was the AFL's drafting of black athletes that brought football into the modern era.[31]

While today football leagues around the nation are integrated racially, the NFL has a long history of institutional racism that continues in the modern era. In 2020 Dan Snyder, the owner of the Washington Commanders dropped the team's name "Redskins" and would later replace it with "Commanders." In addition, Snyder also dropped the Native American caricature which was widely considered an offensive and racist image. That it took until 2020 for a league owner to finally comprehend racist and inflammatory language and images is not just a reflection of Dan Snyder, but rather a reflection of American society at large. In 2021, John Gruden resigned as head coach of the Los Angeles Raiders, walking away from the remaining 6 years and $60 million remaining on his contract. He resigned after some of his emails with racist, misogynist, and homophobic language were leaked to the public. When the leadership of an organization exposes itself as racist, homophobic, or possessing any type of bias, it can hurt the organization's tactical, operational, and strategic performance.

Fritz Pollard, brother of Frank Pollard, both of whom were discussed in Chapter 2 followed his military service as the first African American head coach in the 1920s, serving as a player-coach for Hammond Pros. The second African American head coach was former Oakland Raider offensive lineman Art Shell who was hired in 1989; six decades later.[32] Without Fritz Pollard, the N.F.L. was the last major American sports league to hire an African American to a head coaching position.

While the percentage and numbers have changed on the field, the number of owners, general managers, and head coaches remains dismal. In 2020, the NFL had three (out of 32) black head coaches.[33] Indeed, the NFL recognized this shortfall and instituted the Rooney Rule in 2003. This rule mandated that NFL teams looking to fill a head coaching vacancy must interview ethnic minorities. This rule was expanded in 2020 and now requires the interview of at least two minority candidates for the head coach position, and at least one for any coordinator (offense, defense, special teams) job. Further, the rule expanded to include interviews of minorities for organizational positions such as general manager, and women for senior-level positions.[34] At the owner level, only two NFL owners represent minorities, Shahid Rafiq Khan, a Pakistani-born American, and Buffalo Bills co-owner Kim Pegula who is of South Korean descent.[35]

At the college level, football remained largely a sport of white students through the post-World War II era. It was the University of California Los Angeles (UCLA) that was the first to field multiple black athletes. In 1939, UCLA had five black players on the team including Jackie Robinson, who was a three-sport athlete. Of course, Jackie Robinson would go on to break the color barrier in Major League Baseball a few years later. Despite Robinson's presence at UCLA, many colleges and universities simply refused to play black athletes through the civil rights era. Notably, teams in the Southeastern Conference (SEC) limited their schedules to other SEC teams to avoid racial integration on the field. Sadly, this would not end until 1972 when all football teams in the SEC came to the realization that they simply not compete with other conferences that fielded integrated teams.[36] Diversity is a competitive advantage.

Interestingly, it was head coach Barry Switzer who led the way on racial integration in the SEC. Leading the Sooners of Oklahoma, Switzer understood the competitive advantage of recruiting the best athletes no matter the color of their skin while schools such as the University of Texas were constraining themselves by only recruiting white athletes.[37] Switzer's Sooners would win back-to-back NCAA titles in 1974 and 1975 and another title a decade later in 1985. Switzer would later coach the Dallas Cowboys to a Super Bowl Championship following the 1995 season.

Just as diversity is a competitive advantage, so too is education. Militaries and football teams alike reap the benefits of a force or team that can think. In his book *The Rise of Gridiron University*, Brian Ingrassia details how football became a driving force for universities to attract students. Schools such as the University of Chicago used football as a method to gain publicity and make the American population aware of opportunities for higher education.[38] Universities, to include Harvard and the University of Chicago then became the first institutions to construct large stadiums to host football games and other sports events. In the case of the University of Chicago, the school dropped the football program and eventually turned the stadium into an atomic research facility with a staff led by Enrico Fermi.[39] The work at this former football stadium would contribute to victory in the Pacific Theater in the Second World War.

By in large women have not been a part of football at any level. There is some participation at the youth and high-school levels, and every year or so there is a story of a woman place-kicker on a college team. No woman has played in the National Football League. However, women are making more and more inroads into the broader aspects of the game. In 2015 Sarah Thomas became the first woman to serve as a referee and in 2021 she became the first woman to officiate in a Super Bowl.[40] Jen Welter became the first woman to coach on an NFL team, the 49ers team that made it to the Super Bowl, only to fall to the Kansas City Chiefs. As of 2021, there are eight women coaches in the league.[41]

And of course, in 1997 Amy Trask became the first woman to serve as a team's chief executive, doing so for the Oakland Raiders, a journey she describes in her book *You Negotiate Like a Girl: Reflections on a Career in the National Football League*.[42] Before women served as CEOs in the front office, they appeared in the owner's box. Most famous among them is Georgia Frontiere who owned the Rams after inheriting the team upon her husband's death. Frontiere was the second woman to own a NFL franchise, the first being Violet Bidwell Wolfner who owned the Chicago/St. Louis Cardinals. Wolfner also inherited the team from her husband in 1947 and retained ownership until her death in 1962.

On the field, there is a history of women's professional football. The early 1970s saw the rise of a slew of women's barnstorming teams that led to the establishment of the National Women's Football League (NWFL). This league had several teams that came in and out of existence such as the Dallas Bluebonnets, the Detroit Demons, the Toledo Troopers, the Los Angeles Dandelions, and the Lawton Tornadoes. The league folded in 1979, but today the legacy of the women who played in this league continues with various women's leagues that come into and out of existence.[43] The National Women's Football Association began its first season in 2001 but folded following the 2008 season. The Extreme Football League began shortly after the NWFA disbanded, but is more of a mockery, having its roots in the Lingerie Football League.

It was women's football that attempted to tie in the military community and the football community. In their book *Hail Mary: The Rise and Fall of the National Woman's Football League*, authors Britni de la Cretaz and Lyndsey D'Arcangelo describe how the owner of the Lawton Tornadoes positioned his team adjacent to Fort Sill, Oklahoma, hoping a military community would support the football start-up.[44] Forming professional teams near military bases is risky, as deployments, training cycles, and the whims of the Defense Department's base realignment and closure (BRAC) initiatives often determine the availability of service members to attend games.

More recently, the gridiron has become a political flashpoint. When players in the National Football League, later followed by players at all levels kneeled in silence during the playing of the *National Anthem* to protest police brutality against African Americans, it drew the ire and comments from political leaders at the national level, including then President Trump. Kneeling during the anthem created a division between players and ownership which on a micro-level was an indication of broader political and racial divisions in America.

Both football and the military offer microcosms of American society and life. From the slow and steady advancement of civil rights and equality to the continuous cracks in the glass ceiling through the tenuous relationships between those in positions of power. Football games and the military are not escapes from everyday life with their trials and tribulations; rather they reflect those daily eternal struggles. Even more importantly, both the military and football offer a history to remind the current men and women who fill the ranks and rosters that they stand on the shoulders of those who fought to get them there.

Notes

1. Eliot A. Cohen, *Supreme Command: Soldiers, Statesmen, and Leadership in Wartime* (New York, NY: Anchor Books, 2002), 4.

2. Cohen, 8.
3. Cohen, 209.
4. Dale R. Herspring, The Pentagon and the Presidency: Civil-Military Relations from FDR to George W. Bush. (Lawrence, KS: University Press of Kansas, 2005), 44–45.
5. Ibid.
6. Cohen, 191.
7. Cohen, 192.
8. Bob Woodward, *Obama's Wars*. (New York: Simon and Schuster, 2010), 372–373.
9. Jonathan Alter, "Secrets From Inside the Obama War Room." *Newsweek* Web, 31 May 2010.
10. Marybeth P. Ulrich, "A Primer on Civil-Military Relations for Senior Leaders." *In U.S. Army War College Guide to National Security Issues, Vol. II*, J. Boone Bartholomees Jr., ed., (Carlisle Barracks, PA: U.S. Army War College, June 2008), 665.
11. Williamson Murray, *American and the Future of War: The Past as a Prologue*. (Stanford, CA: Hoover Institution Press, 2017), Kindle.
12. David Halberstam, *The Education of a Coach*. (New York, NY: Hachette Book, 2012).
13. Phillips Payson O'Brian, *How the War was Won: Air-Sea Power and Allied Victory in World War II*. (Cambridge, UK: Cambridge University Press, 2015), Kindle.
14. Andrew Roberts, *Masters and Commanders: How Four Titans Won the War in the West, 1941–1945*. (New York, NY: Harper Publishing, 2009), 341.
15. Former football players are overwhelmingly men. It is only recently that women have broken into the assistant coaching ranks on NFL teams. As of 2021, every player, head coach, and offensive or defensive coordinator has been a man.
16. End strength is the size of the force, or the number of people in uniform.
17. Catherine Lutz, *Homefront: A Military City and the American 20th Century*, (Boston: Beacon Press, 2001), 121.
18. Kate Werren, *An American Uprising In Second World War England: Mutiny in the Duchy*. (Yorkshire, UK: Pen and Sword Books, 2020).
19. Marvin E. Fletcher, *America's First Black General: Benjamin O. Davis Sr., 1880–1970*. (Lawrence, KS: University Press of Kansas, 1989), 84.
20. Doug Melville, *Invisible Generals: Rediscovering Family Legacy, and a Quest to Honor America's First Black Generals*. (New York, NY: Atria/Black Privilege Publishing, 2023), 33.
21. Beth Bailey, *An Army Afire: How the US Army Confronted Its Racial Crisis in the Vietnam Era*. (Chapel Hill: University of North Carolina Press, 2023).
22. David F. Burrelli, "The Repeal of "Don't Ask, Don't Tell": Issues for Congress." Congressional Research Service. December 21, 2012. Retrieved from: https://fas.org/sgp/crs/misc/R42003.pdf
23. Amanda Boczar, "Managing Harassment and Assault in the Contemporary U.S. Military" in *Managing Sex in the U.S. Military: Gender, Identity, and Behavior*, ed. Beth Baily, Alesha E. Doan, Shannon Portillo, and Kara Dixon Vuic. (Lincoln, NE: University of Nebraska Press, 2022), 230.
24. David McCullough, *Truman*. (New York: Simon and Schuster, 1992).
25. Herspring, 68.
26. Herspring, 70.
27. John Eisenberg, *The League, How Five Rivals Created the NFL and Launched a Sports Empire* (New York, NY: Basic Books, 2018).
28. Ibid.
29. Matthew Sherry, *Any Given Sunday, The NFL's Epic 100-year History in 20 Games*. (London, UK: Orion, 2020), Kindle.
30. Michael Oriard, *Brand NFL: Making and Selling America's Favorite Sport*. (Chapel Hill: University of North Carolina Press, 2007), 259.
31. Ian S. Kahnanowitz, *NFL 1970: The Inaugural Season of the New NFL*. (Mechanicsburg, PA: Sunbury Press, 2021), 32–33.

32. Michael Janofsky, "Shell is First Black Coach in N.F. L. Since 20's." *The New York Times.* 4 October 1989. Retrieved from: https://www.nytimes.com/1989/10/04/sports/shell-is-first-black-coach-in-nfl-since-20-s.html
33. 1) Mike Tomlin of the Pittsburgh Steelers, 2) Brian Flores of the Miami Dolphins, and 3) Anthony Lynn of the Los Angeles Chargers.
34. Kevin Patra, "NFL instituting changes to Rooney Rule." NFL.com 18 May 2020. Retrieved from: https://www.nfl.com/news/nfl-instituting-changes-to-rooney-rule
35. Kevin Patra, "NFL instituting changes to Rooney Rule." NFL.com 18 May 2020. Retrieved from: https://www.nfl.com/news/nfl-instituting-changes-to-rooney-rule
36. Paul, Joan, Richard V. McGhee, and Helen Fant. "The Arrival and Ascendence of Black Athletes in the Southeastern Conference, 1966–1980." *Phylon (1960-)* 45, no. 4 (1984): 284–97. Accessed July 19, 2021. http://doi/10.2307/274909.

    Kentucky was the first school to allow African-American football players to compete in 1967, Ole Miss was the last team to do so in 1972. It was not until 2003 when Ole Miss retired their racist mascot, Colonel Red, and 2015 when the Rebel flag was removed from the campus.
37. Tim Layden, *Blood, Sweat, & Chalk: How the Geniuses of Football Created America's Favorite Game.* (New York: Sports Illustrated Books, 2010).
38. Brian M. Ingrassia, *The Rise of Gridiron University: Higher Education's Uneasy Alliance with Big-Time Football.* (Lawrence, KS: University of Kansas Press, 2015).
39. Ibid.
40. Andy Gallagher, "Who is Sarah Thomas? Super Bowl 55 female referee making history." 7 February 2021. Retrieved from: https://www.profootballnetwork.com/sarah-thomas-nfl-female-official/
41. Gillian R. Brassil and Kevin Draper, "These Women Were N.F.L. "Firsts." They're Eager for Company. *New York Times.* 3 February 2021. Retrieved from: https://www.nytimes.com/2021/02/03/sports/football/nfl-women-coaches.html
42. Amy Trask and Mike Freeman, *You Negotiate Like a Girl: Reflections on a Career in the National Football League.* (Chicago, IL: Triumph Books, 2016). Kindle.
43. de la Cretaz and D'Arcangelo,
44. de la Cretaz and D'Arcangelo,

Chapter 12
The Similarities

Practice doesn't make perfect. Perfect practice makes perfect. –Bill Walsh.

Give me enough medals and I'll win you any war. – Napoleon Bonaparte.

We're ten and one! – Bill Murray as John Winger in Stripes.

Tactics are the level of individual battles and engagements. Football, and more broadly sports in general teach us individual life lessons. From childhood through adulthood, participation in sports and athletic activities develops and helps to define who we are. Youth sports teach us discipline, how to interact with others as part of a team, and how to solve problems on our own. Whether it's football, baseball, tennis, basketball, or hockey, sports provide an arena to practice the skills it takes to participate and lead in the military.

One of the most particular aspects of war and football comes from Bill Parcels, the former head coach of the New York Giants, New England Patriots, New York Jets, and Dallas Cowboys who once quipped "you are what your record says you are."[1] There are multiple levels to this insight.

Football teams generally realize who they are about a quarter of the way through a season, and at the halfway point, teams that lost more games than they won understand where they stand, and coaches, general managers, and owners must decide if the team needs to move in a different direction. Teams with healthy winning records know who they are, and generally begin to look forward to competing in the postseason.

A nation can be boastful about the power of its military, but sooner or later said nation must look at the record. America for example, began with a winning streak lasting for nearly 170 years. The winning streak then turned into an undefeated streak with the stalemate on the Korean Peninsula, followed by its first major loss in Vietnam. The battlefield victories in Grenada, Panama, and the 1991 Gulf War seemed to turn the tide back in America's favor. However, 19 years of fighting endless wars in Afghanistan, Iraq, and Syria force one to examine and think about our record.

"You are what your record says you are" applies not just to the strategic outcomes of a nation but is equally applicable to battlefield commanders at the tactical level. Colonels and generals commanding brigades and divisions should stand by their

D. Sukman, *American Football and the American Way of War*,
https://doi.org/10.1007/978-3-031-55345-5_12

battlefield performances, and the performance of their subordinate units. Unfortunately, as retired Colonel Paul Yingling pointed out in 2007, the general officer corps of the U.S. Army was never held to account for their losing record.[2] Indeed, according to Yingling, the general officer corps of the U.S. military failed to develop and implement a proper strategy to stabilize Iraq and defeat its respective insurgency. You are what your record says you are, and decades of fumbling around in Iraq, Afghanistan, and other areas in the broader Middle East speak for itself.

The Centrality of Violence: In 1905 eighteen people died from the injuries they sustained on the football field.[3] In 2020, ten American service members died while deployed to Afghanistan.[4] While these two statistics represent the extremes on the spectrum of violence in each activity it is clear that in both war and football are defined by what society considers an acceptable level of violence against fellow human beings.

It is on the gridiron where many of the youth in America experience violence and hitting other people for the first time. While the game is not intended to injure people, and the goal of each player is not to injure another, serious injuries from concussions to broken bones to torn ligaments are a constant reminder of the dangers of the game. As players move from one level to the next, the size of their fellow athletes increases as does the speed of the game. As force = mass x velocity, football isn't a contact sport, it's a collision sport with violence as the defining feature (Fig. 12.1).

In October of 1965, Thomas B. Morgan wrote in Esquire Magazine what is perhaps to this day the defining description of football's violence as a comparison to war.[5] Players on the gridiron know and understand that inflicting pain on the opponent is a valid tactic to achieve victory, just as attrition is a valid strategy to pursue in war. But just as war is more than individual battles between soldiers, football is more than one player seeking to hurt an opposing player standing before them. Football, like war, is often about one team causing more pain to the opposing team.[6]

On the battlefield, committing acts of violence against an enemy is one of the ways to achieve victory, tactically, operationally, and strategically. Indeed, wars are often won when one side can no longer sustain the casualties they suffer. While the laws of land warfare have evolved to limit what one soldier may do to another, killing other human beings remains central to war.

Acceptable levels of failure: Anyone who claims to have never failed at anything is probably trying to sell you something. Learning to fail on the sports field is in fact, one of the most valuable life lessons one can learn. Everyone needs to experience failure at some level throughout his or her lifetime. Some failures you can recover from, others will haunt you for the remainder of your life.

There is a difference between a 3-yard loss on first down and giving up a 100-yard kickoff return. Sports prepares us to deal with failure. Not only to accept certain levels, but to overcome failures with a better performance on the next down, the next series, or the next game. Everyone fails at some point in his or her military career. There are many layers to failure in the military. It can be as simple as blowing a brief to a commander or getting recycled at ranger school. We can bounce back from some failures, but a failure that involves compromising morals or ethics can

Fig. 12.1 Cover of October 1965 Esquire magazine

destroy a career. Sports also teaches us that there are people who help us overcome our failures along the way. Coaches and teammates from athletic programs translate to commanders and peers in our military careers.

Reliance on Others*:* Often, youth sports, and more specifically youth football, is the first experience of being a part of a cohesive group of individuals over a long period. While coaches may change out from junior high school to junior varsity and varsity, the players on each team by in large remain the same. Youth football often has a core group of players who began playing at the age of eight through the end of high school. The bonds developed in childhood last a lifetime.

Similarly, in the military, service members may be a part of a unit that spends over a year preparing for, and then a year serving in combat. The importance of team chemistry and developing relationships that go beyond the workplace is paramount to warfighting units. At the tactical level for example, soldiers build fighting positions with interlocking fields of fire. This method means that it is the responsibility of soldiers in the foxhole next to them to fire at the enemy in the dead space directly in front of them.

Football teaches us that success and failure is a team effort. Miss a block or forget an assignment in pass coverage and the opponent will capitalize on the error. Even worse, a missed assignment can lead to a debilitating injury to a teammate. In the military, we learn quickly that success is a team effort. Teamwork can mean knocking down targets in an assigned sector of fire or contributing ideas within an operational planning team. Indeed, failure at the individual level can lead to mission failure, or even worse getting someone close to you killed. Athletes, just like soldiers, sailors, airmen, and marines must be willing to take care of their responsibilities and trust that their teammates will do the same.

Next Man (or Woman) Up: Sports, as war, is a young person's game. In every team sport, athletes get sick, athletes are injured, or have a personal issue that makes them unavailable to compete. When this occurs, the backups or the second-string steps in to take the place of a starter. Moreover, there are times when the potential of a backup player becomes more valuable to a team than the value of the current starter. Indeed, there are multiple stories of backup players leading their respective teams to glory. For example, in 1990, Jeff Hostetler replaced an injured Phil Simms at quarterback and led his team to an upset win over the Buffalo Bills in Super Bowl XXV. In 2001, the Patriots back-up quarterback Tom Brady stepped in for Drew Bledsoe following a vicious hit from the Jets' Mo Lewis. Twenty years later, Tom Brady is widely regarded as the greatest quarterback to play the game, but that greatness was not enough to overcome Nick Foles, the Philadelphia Eagles backup quarterback in Super Bowl LII. The lesson this teaches us is that you never know when your opportunity comes. When it does, make the most of it.

In war, soldiers get hurt and sometimes soldiers die. Other times soldiers need the time to take leave for the birth of a child or to take care of a sickness in the family. Just as in football (or hockey, soccer, or any other sport), another soldier must step up and assume the duties and responsibilities of the soldier who is gone. Indeed, the concept of war fought through attrition means that militaries, just like football teams must develop depth in their ranks, and the capability to replace the dead and wounded. It is a common practice in many military units to have soldiers learn their job and one job up, so that are ready to step in should casualties occur.

Military units and football teams understand that a system is more important than the individual. If one person in a squad, platoon, or on offense or defense is indispensable the organization is brittle and ripe for failure.

How you practice is how you play and How you train is how you fight: The ability for the American military to provide realistic training is an asymmetric advantage we hold over our adversaries. From the time one competes in youth sports, this idea is drilled into athletes in childhood. The amount of effort teams put forth on the practice field easily translates to success or failure on game day. Over the past 16 plus years, the U.S. has been witness to this concept in war. Indeed, during the first Gulf War in 1991, tank crews stated that rotations at the National Training Center (NTC) were more difficult than actual battle. Hard, realistic training at home station and at NTC and the Joint Readiness Training Center (JRTC) enable better combat performance for tactical level units. High levels of mental preparation and planning in events such as Combatant Command exercises and division or corps level warfighter exercises are critical for military success in times of crisis. In both sports and the military, the more effort you put in on the front end, the better the chance of success on the field.

Football teams and military units train with a similar philosophy of concentrating on individual level skills and gradually working up to collective training, culminating in full team or unit exercises. For example, at the individual level soldiers train in first aid, rifle marksmanship, and other job specific skills. Individual training leads to squad, platoon, company, and later battalion and brigade level training events. Training tends to culminate in force-on-force exercises at Combat Training Centers (CTCs) such as Fort Johnson's Joint Readiness Training Center, and Fort Irwin's National Training Center. The Air Force conducts force on force exercises in Nevada in an event known as Red Flag.

Like the military football teams have their players practice their individual skills and conditioning based on their position. The individual skills then move to collective drills where the offensive line or receivers and quarterbacks work together. Practices then move to intra-squad scrimmages where the offense and defense line up against each other. In the pre-season teams often schedule scrimmages against other teams in the league.

Chemistry Matters: Chemistry, according to Sam Walker, the author of *The Captain Class* is "the interpersonal dynamics that have an impact on team performance."[7] In football, a combination of the system and talent will only go so far. How players on the field interact is paramount to success. Described as chemistry, some players develop a deep rapport with their teammates and rapidly develop an instinct to know where their teammates are on the field. Quarterbacks must be in synch with their receivers to deliver the football to the right place at the right time.

From tactical level units such as an infantry squad and a tank platoon, to Operational Planning Teams (OPTs) solving strategic problems, chemistry is vital. At the tactical level, a high level of chemistry can enhance a unit's ability to execute basic battle drills, A tank crew must develop trust with each other as they operate a 70-ton machine designed to kill people. Moreover, the chemistry between units enables actions ranging from combined arms maneuver to joint and coalition warfare.

At the operational and strategic levels of war, chemistry often comes with continuity. Teams that have predictable leadership tend to perform better over the long run, while teams that constantly change leadership tend to suffer with weak performances. In football, this becomes clear with teams that stick to a head coach and staff. Over the long term, teams such as the New England Patriots (Bill Belichick) and Pittsburgh Steelers (Chuck Noll, Bill Cower, Mike Tomlin) who keep coaching staffs in place consistently outperform teams such as the Cleveland Browns who historically been quick to pull the trigger on a head coach after one losing season.

Hampering the performance of the U.S. Military in both Iraq and Afghanistan was the constant rotation of forces and leadership. America fought for eight years in Iraq (2003–2011), and as of this writing is entering its twentieth year of war in Afghanistan. However, often there is an argument that we didn't fight in Iraq for eight years, rather we fought for one year eight consecutive times. Ditto for our time in Afghanistan. In both theaters of war, new commanders at the tactical and operational levels deployed into theater with their own concepts and plans to win the war.[8] This aspect created friction in our dealings with the political and military leadership and with the populations of each respective country.

Intelligence Preparation of the Operational Environment: When athletes begin to compete, numerous hours each week are spent in the film room. These hours include a review of past performance and a scouting of the next opponent. Players and coaches take the time to understand their formations, the tendencies of their adversary, and what plays would most likely be called in each situation. To an athlete entering the military, this is the first exposure to an enemy doctrine template. In a similar manner, teams take the time to understand the field they will play on and the expected weather on game day…a situation template. Sun Tzu, in his book *The Art of War* pressed the imperative that to be successful, you must "know thy enemy, and know thy self."[9] A deep understanding of an opponent's capabilities and most likely course of action allows a force or a sports team to focus its preparation. However, understanding that the opposing force is conducting the same preparation, watching your team on film, or studying your doctrine leads to the next way sports prepared me for a military career.

A living, thinking, and adapting adversary: War is the most complicated of human endeavors and in both sports and war, the decisive advantage goes to the team or force that learns faster than the other does. Relying on the weakness or incompetence of an enemy is a key ingredient to failure. A key aspect of the difficulties of warfare is that no nation or army conducts operations against an inanimate object. Enemies and adversaries constantly adapt to the environment while doing their best to defeat you. Adversaries have strategies and adjust them on the fly. Those who fail to adapt often face defeat. This description fits both football and war.

Athletes learn early in life that no matter how well-prepared a team may be on the practice field, or in the gym, opponents are doing the same thing. Planning and working to win by putting in time at the gym to out-muscle their respective opponent on the field. They adjust their game plan for every series, hoping to learn faster than their opponent does. Defining success is who can stay one step ahead. In a similar plane, understanding that adversaries adapt leads to the wisdom of never underestimating an opponent.

The system, not the individual: In football, no single player is greater than the collective system. In a team sport such as basketball, where the best player can be on the court for most of the game, the teams with the best players in the game tend to rise to the top (e.g. the 2016 Cavaliers and Warriors). However, in football, the best receiver may only touch the ball fewer than ten times in the game. When you include downs on special teams, the best offensive and defensive player may be on the field for less than half the game. The best teams play a system that highlights the strengths of its players. Team systems triumph over individual talent.

The U.S. Army is filled with thousands of smart, talented, and motivated people. And every single one of these people, from the newest private to the Chairman of the Joint Chiefs is replaceable. Where football teams employ systems such as the Run and Shoot offense, the 4–3, and 3–4 defense, the military has a wide range of "systems," ranging from doctrine to weapons platforms. Any one person can step into a position they have been trained for and adapt to the specific skills required for the position. Whether a soldier is a squad leader or joint planner on the first day of the job, there is doctrine to read and fall in on.

Within the concept of systems, football and the military have one further similarity. The ability to leverage creativity and innovation within a system. From a football perspective, a team can divert from the system to take advantage of a turnover by an opponent, or lucky bounce of the ball. From the military aspect, doctrine is a guide to tactics and operations, not gospel. Leaders at all levels have the capability to diverge from doctrine when the situation demands it.

Think fast act fast: This is where the ideas of Colonel John Boyd that were discussed in the strategic theory chapter come to light. Football players must make decisions on what they will do with the ball before they throw, receive, or defend a pass. The time to execute a pass or to make a skilled move with the ball can be less than one second. Indeed, quarterbacks with the quickest release time hold an advantage over the defense. Quarterbacks must rapidly scan the field for an open receiver and release the ball in under three seconds before they feel the impact of a 300-pound defensive lineman running at full speed.

As players move up in each level of football the speed of the game gets faster. But the speed of the game is more than just a player's 40-yard sprint time, it's the speed of decision making. Running back must decide to turn up field faster, defensive backs must decide when to break on the ball faster, and quarterbacks must make their reads and decide where to throw the ball at a pace that no other level of the game can match.

At the tactical level of warfare, decisions on the use of force, or when and how to engage a target occur in a matter of seconds. This is true for the soldier standing guard at a checkpoint, or for a tank commander identifying a target, to counter-battery fire. Having situational awareness, an understanding of the environment, and trust in the ability of teammates allows for rapid decision making, be it on the gridiron or in the desert.

As one goes up in the levels of war from tactical to strategic, decisions have more impact and assume greater risk. This holds for operational, strategic, and institutional decisions. For example, committing an operational level reserve force to reinforce success or prevent failure of an operation is a critical decision of battlefield commanders. Once committed, this force is unavailable should they be needed somewhere else. Institutionally, decisions on acquisitions on major combat platforms cost billions of dollars and result in the fielding of systems that remain in the force for decades. Indeed, the M1 Abrams Tank and the B-52 Bomber have had multiple generations of tankers and pilots riding them into battle.

Operational, strategic, and institutional decisions for a football team often outweigh the impact of the tactical level decisions during a game. Draft picks can make or break an organization for years. Getting the right personnel on the team has a direct influence on the long-term performance and competitiveness of an organization in a sport will billions of dollars are at stake. Understanding the impact of decisions should lead to more thoughtfulness, and if necessary, time allocated to the decision.

Be thick skinned: Every player of a visiting or away team will at some point or another be told how much he "sucks" by fans of an opposing team. Often, the intent of riling up an opposing player is to get them out of their comfort zone, to get them to lose control of their emotions, and in the end to take an unnecessary penalty at an inopportune time. The best players can tune out the insults and focus on their game. A second aspect of thick skin is the ability to overcome mistakes. The smallest error or miscue can rapidly become a viral meme on the internet, or a lowlight in the next day's *Sports Center* broadcast. Football players must cultivate an internal ability to ignore this noise can focus on the next play, game, or season.

The ability to control emotions is essential for anyone in the military. Emotional control can range from enduring an "ass chewing" from an angry boss, to remaining calm and collected under fire, and all the stressful events in-between. Service members in leadership positions must be able to lead their platoons, battalions, or operational planning teams in times of great stress.

A wide range of talent: From the lowest age groups in Pop Warner Football leagues to the highest level of professional athletes, there is a pecking order of talent. Levels of talent manifest in multiple ways. This includes who is in the starting lineup in youth leagues to who receives the largest paycheck in the NFL. Early in life, the top tier talent is often a result of who is the biggest player on the field, which itself is often the consequence of what month players are born. Older kids tend to fair better than younger kids.

Much like a football team, within every job or military occupational specialty (MOS) in the military, there are various layers of talent. If you think of yourself as the most fit infantryman, or most intellectual strategist, chances are there is someone who can outrun or out-think you. Athletes and soldiers must continually enhance and refine their skills, whether it's a 40-yeard dash run time, or the ability to hit every target on the rifle range.

Understanding the talent within an organization is a driving factor of success in both football and war. In recent years, the U.S. Army took steps to modernize its talent management system at the strategic level. Indeed, beginning in 2020, the Army instituted a Battalion Commander Assessment Program (BCAP), modeled after the NFL's pre-draft combines. Moreover, the Army is expanding on this idea and will experiment with a Brigade Commander Assessment program beginning in

2021. These programs measure the physical fitness, psychological readiness, and leadership qualities of top tier officers competing for a limited number of command assignments.

Judging the talent of an individual, and how they will perform on the football field and on the battlefield is not an exact science. Donn Starry, a former 4-Star General once quipped that "the battlefield is the most honest place on earth."[10] War exposes who can lead, and who can't. War quickly determines who is competent and incompetent. We can do our best to train and test military personnel in leadership but can never truly predict how anyone can respond to the carnage of warfare.

Football coaches and general managers judge the talent of prospective players through performance in previous levels of the game. For example, college coaches spend countless hours reviewing films of high school athletes, and NFL coaches do the same for college players. Information gleaned from the film is then combined with pre-draft workouts, interviews, and psychological examinations.

Technology matters, but it's not everything: Every team, at nearly every level, has a player who owns the most high-tech equipment. This can include $1000 cleats, the most high-tech pads, and the application of the latest in sports science for health and fitness. While these technologies may improve a player's confidence, and to an extent may improve performance, the equipment does not make the athlete. Talent does.

Military forces across the globe consistently pursue the latest in equipment. Indeed, the American way of warfare often relies on the latest in technology. The second and now third offsets rely on the United States holding distinct advantages over foreign adversaries in the development and fielding of technology. Examples include precision guided munitions, stealth aircraft, and long-range fires. However, talent still matters. As witnessed in Vietnam, and more recently the wars in Iraq and Afghanistan, technology is not a guarantee of victory. Other factors such as training, education, and leadership still hold sway in combat.

The critical aspect of applying or adapting new technology either to a military or to a sports organization is taking the time to evaluate, test, and validate that the new technology will solve specific problems. In football and the military acquiring new and fancy technology without a purpose can risk confusing those who use it, or diverting attention, money, and time away from other more relevant training. Further, it takes time to train soldiers and athletes alike on how to best use new technology before using it at the expense of already proven capabilities.

The Value of Awards: There are few honors as respected and valued in American society as the Medal of Honor. It is the highest award for bravery and valor a service member can receive. However, the Medal of Honor is one of dozens of medals and ribbons military service members can earn over the course of a military career. Other combat awards for valor and service include the Sliver Star, the Bronze Star, Commendation and Achievement Medals, the Purple Heart, the Distinguished Service Cross, the Air Medal, and the Legion of Merit. Moreover, the U.S. military awards campaign medals and ribbons for various missions. After 9/11 the military awarded the Global War on Terror Service and Expeditionary Medals, medals for serving in Iraq and Afghanistan, allowing service members to wear NATO medals for service under a NATO headquarters in Afghanistan or the Balkans. Medals are an important facet of military service as promotions happen on detailed career timelines and there are no financial bonuses for individual acts of valor.

Football leagues at all levels use their own awards systems to incentivize better and higher performance from players and coaches. Team and league Most Valuable Player (MVP) often come with a trophy and recognition from peers. Annually, college football awards the Heisman Trophy to the best player among other awards.[11] Other awards include offensive and defensive player of the week and player of the month, rookie of the year, most improved or comeback player of the year, and coach of the year. Division, conference, and league championships often come with a trophy for the team and championship rings for the individual players. Teams also hang banners in their stadiums to remind fans of past accomplishments. At the professional level NFL players and coaches have financial incentives in their contracts to achieve these recognitions and to attain individual statistical accomplishments (rushing yards, sacks, etc.). Both the military and football organizations understand the value of bling as an incentive for better performance.

Mentorship: Throughout one's life, there are a select few individuals who serve as mentors. This group includes parents, teachers, and coaches. To a child playing youth football, the coach assumes an outsized role where his or her word is the law on the field. Players listen attentively to every word and direction of the coach, understanding that failing to do so would result in less playing time during a game.

Like coaches, leadership and mentorship are foundational to the military profession. Mentorship is responsible for the rise of some of the greatest military commanders to emerge in the 20th Century. General Fox Conner for example is known to have mentored the likes of Generals George S. Patton, George C. Marshall, and Dwight D. Eisenhower.[12]Indeed, the First World War served as a mentorship factory for the men that would lead the U.S. military twenty years later in World War II. Climbing the career ladder in any profession is difficult without a guide or a someone to champion one's knowledge, skills, and abilities to those in power. At the highest levels, fair or unfair, mentorship can be the decisive advantage.

Rivalries: As of 2021, the United States has five stated adversaries or competitors, depending on how you classify them. *The 2017 National Security Strategy* and subsequent administration's *2022 National Security Strategy* along with their respective subordinated nested strategies such as the National Defense Strategy and National Military Strategy call out China, Russia, North Korea, Iran, and Violent Extremist Organizations (VEOs) as our opposition to attaining American national interests. The actions of these nations and VEOs run counter to what America seeks to accomplish. For example, arguably the construction of China's Belt Road comes at a cost to American influence and economic prosperity. Further, Russia's behavior in cyberspace and interference in U.S. elections stands as a threat to American democracy.

The stated adversaries did not suddenly appear out of nowhere in 2017. America's competition with Russia dates to the end of World War II and the Cold War competition with the Soviet Union. Iran has been an enemy of the United States since their

Islamic Revolution in 1979, and North Korea has been an enemy since their army crossed into South Korea, leading to American Military action in the Korean War over seventy years ago.

Professional and amateur sports, and football are defined by rivalries. The evolution of rivalries tends to relate to geographic location. Pop Warner and high school football teams often have rivalries with neighboring towns. College athletics are arranged in broader geographic regions such as the South Eastern Conference (SEC) and the Western Athletic Conference (WAC). At the professional level, the NFL organizes divisions based on geography. (e.g. NFC East, NFC South, NFC West). Organizing leagues in this manner ensures that the same teams play each other year after year. Coaches, although judged on their win-loss records, also gain a reputation based upon their team's performance against their chief rivals. A one-loss season normally considered a success can be a failure for an Alabama head coach whose only loss comes to Auburn in the annual Iron Bowl. Conversely, a West Point team can salvage a losing season with a decisive win over Annapolis at the annual Army-Navy game.

Hurry Up and Wait: The life of a soldier in combat is often characterized by short and intense moments of combat that occur between long periods of intense boredom. The phrase "hurry up and wait" is a common description of military life both in garrison and in a theater of war. Indeed, the U.S. military invests millions of dollars for morale, welfare, and recreation activities at home and abroad. Any soldier who has spent a day on a rifle range waiting to qualify or sitting in a motor pool waiting for an accountability inspection understands this aspect of the military profession.

George Plimpton once described football as, "violence punctuated by committee meetings."[13] There is no better description of life in the military or a football game than this quote. Like war, a football game occurs in short spurts of activity between longer periods of relative inactivity. According to the authors of *The Economics of the Super Bowl*, while a football game might take 3–4 hours to complete, games at the professional and collegiate level only have 60 minutes of playing time, of which players are engaged with the opposing team for about 11 minutes.[14] Huddles between plays, timeouts, injury timeouts, transitions between offense, defense, and special teams and transitions between quarters and halftime account for most of the game. At the higher collegiate and professional level, time is allocated for television time outs to maintain revenue.

There is a life after football, there is a life after the military: Sports and war are a young person's game. Most professional sports players are in their 20 s, while most sports players in America are those in youth, high school, and college programs. The average career in the National Sports League spans less than four years. Unlike the game of chess, or the lifelong sports of tennis and golf, sooner or later, everyone's sports playing days are over. This aspect is the same for those who choose military service. Soldiers tend to be young and in their 20 s, and even those who pursue a 20-year career are looking for another job by the time their early 40 s.

The U.S. military operates on the "up or out" promotion system. Officers have a window for selection to the next rank, and should they miss that window, it's time to search for a new job. To continue to advance in rank officers are rated against their peers year after year. In a way, this constant competition is similar to how football players are always competing for a starting job over serving as a backup and fighting to stay on the team as a backup over getting cut. In either profession, every year is a rat race where every individual is competing against their teammates, be it in a football uniform or a military uniform.

Nobody ever likes to be told that they have reached the maximum potential of the craft they love. For athletes, this comes when they are cut from the team during training camp, or simply don't get a contract extension or renewal. For service members, it comes when a centralized selection board doesn't choose you for the next level of professional military education, for a command assignment, or for the next promotion. In any of these cases, the effects can be a devastating dose of reality. Sooner or later, there comes a time when you must think about the next phase of life.

Notes

1. Bill Parcells. "Parcellsisms: You are What Your Record says you are. Interview with Nunyo Demasio. 7 October 2016. Retrieved from: https://www.youtube.com/watch?v=a1dtAVNEnkQ.
2. Paul Yingling. "A Failure of Generalship," Armed Forces Journal. 1 May 2007. Retrieved from: http://armedforcesjournal.com/a-failure-in-generalship/
3. John J. Miller, *The Big Scrum: How Teddy Roosevelt Saved Football* (New York, NY: Harper Collins, 2011 Kindle.
4. Phillip Walter Wellman, "Afghanistan: Remembering the fallen of 2020." *Stars and Stripes*. 30 December, 2020. Retrieved from: https://www.stripes.com/news/middle-east/afghanistan-remembering-the-fallen-of-2020-1.656849
5. Thomas B. Morgan, "The American War Game." Esquire Magazine. October, 1965. 71–72, 142, 146–148. https://classic.esquire.com/issue/19651001/print
6. Ibid.
7. Sam Walker, *The Captain Class: A New Theory of Leadership*, (New York, NY: Random House 2017), 35.
8. To see a terrific example of this aspect, watch the Netflix Movie *War Machine*, starring Brad Pitt.
9. *Sun Tzu: The Art of War.* Trans. Samuel B. Griffith. (Oxford University, 1971), 84.
10. Mike Guardia, *Crusader: General Donn Starry and the Army of His Times* (Philadelphia, PA: Casemate Publishers 2018), Pg 168..
11. In addition to the Heisman, other trophies include the Jim Thorpe Award for best defensive back, the Butkus Award for best linebacker, the Biletnikoff Award for best receiver, and the Lou Groza Award for best placekicker.
12. David T. Zabecki, "Brigadier General Fox Conner" in *Pershing's Lieutenants: American Military Leadership in World War I*. ed. David Zabecki and Douglas Mastriano. (Osprey Publishing, 2020). Kindle.
13. Will Carroll, Tyler Brooke, and Peter King, *The Science of Football: The Math, Technology, and Data Behind America's Game*. (New York, NY: Sports Publishing, 2022), 9.
14. Yvan J. Kelly, David Berri, and Victor A. Matheson, *The Economics of the Super Bowl: Players, Performers, and Cities*. (Cham, Switzerland: Palgrave Macmillan), 118.

Chapter 13
The Differences

The battlefield is the most honest place on earth. – General Donn Starry

Pro football doesn't build character, it reveals it. – Marv Levy

War is a Racket – General Smedley D. Butler.

13.1 The Obvious

The previous chapters looked at football as a useful way of explaining the various concepts of international relations, military theory, strategy, and U.S. joint warfighting doctrine. However, one should never mistake a game for war. While this chapter will discuss some of the differences, it will start with the obvious that when a military begins to fight a war, the result is often thousands to millions of dead, maimed, and injured people. The dead come in all categories and include service members from all sides, non-state actors, and unfortunately, and all too often innocent civilians. While the era of satellites, global positioning systems (GPS), and precision-guided munitions (PGM) reduced the massive numbers of civilian casualties seen in the firebombing of Japanese and German cities in World War II, they have not eliminated them. Indeed, the use of PGMs is a luxury of wealthy nations and all too often not available to second and third-tier powers or non-state actors.

While football is a contact sport where players often sustain injuries, some of which can be permanent and life changing, it is not from a deliberate effort of the opposing team. The Chicago Bears despite their long-time rivalry with the Green Bay Packers are not trying to kill players on the other team.

Of course, the purpose of war is for a nation to achieve policy objectives or to maintain national interests when other instruments of national power such as diplomacy, economics, and information fail to do so. These policy objectives might be to eliminate a growing threat, to maintain freedom of navigation on the open seas, or to gain or recapture previously lost territory. Nations typically don't go to war simply to fight a war. Football on the other hand is a sport at the lower levels, and a business at the professional level. The objective is to compete, to have

D. Sukman, *American Football and the American Way of War*,
https://doi.org/10.1007/978-3-031-55345-5_13

Fig. 13.1 Live CNN coverage of the 1991 Gulf War (*Source* CNN)

fun, to entertain, and to make money. This difference notwithstanding, there is an entertainment aspect to modern day conflict (Fig. 13.1).

The modern era consists of a persistent information environment that militaries operate in on the modern battlefield. Satellite and internet technology allow service members, civilians, government officials, and media to provide real-time video and commentary in war. The gradual shift from reporters sending articles back to the newsroom to instant analysis began with the live coverage of the 1991 Gulf War. This continued and led to today's environment where nations and non-state actors use information as a weapon. ISIS for example presented to the world live coverage and uploaded videos to YouTube of their atrocities. This served to intimidate their enemies, but conversely helped to form an international alliance against them. During the 2003 U.S. invasion of Iraq, multiple news organizations embedded reporters with U.S. units that provided live images of battles between U.S. and Iraqi forces. The 2022 Russian invasion of Ukraine led to massive amounts of video clips on multiple social media sites of drones and artillery strikes destroying equipment and killing soldiers. Selling war as entertainment makes for good television ratings and leads to thousands of "likes" on the internet.

It's not just the live pictures that garner ratings, war is an essential part of literature, television, and film. There are World War I and World War II classics such as Remarque's *All Quiet on the Western Front* and Heller's *Catch-22,* which are often read and taught in high schools and universities across America. Tom Clancy novels top best seller lists while war themed television and films range from comedies such as *Hogan's Heroes, Dr. Strangelove,* and *McHale's Navy* to action films such as *The Dirty Dozen, Black Hawk Down, The Outpost*, and the *Rambo* series to dramas such as *Bridge Over the River Kwai, Apocalypse Now*, and *Platoon.* Television too offers war and military themed shows such as *Space Force*, *Tour of Duty*, and *M*A*S*H* whose final episode still ranks as the highest rated episode in television history.

While war certainly crosses into the entertainment realm, football is more than a game. Football in America is a multi-billion-dollar business. Youth sports and high schools across America spend significant portions of their budgets on equipment and facilities for local teams. Colleges and universities offer various levels of scholarships to football players, often serving to fund an education. NCAA college football coaches are among the highest paid personnel on their respective campuses. NFL Football consistently ranks as the highest rated television program each week during the season, generating millions of dollars in advertising revenue. Football is an American cash cow, and if economics is an instrument of national power, football is a key element to that instrument.

While the stakes are not the same, it doesn't mean that football doesn't matter. In his book *Why Football Matters: My Education in the Game*, Mark Edmundson talks about the safe and clearly defined environment in which football games occur.[1] Win or lose, you walk off the field ready to practice and compete the next week (assuming the avoidance of a major injury). But as Edmundson opines, football does matter, to the coaches, the fans, the players, and their families. In the moment what is happening on the field means everything.

13.2 One On One Versus Alliances

Professional football teams fight a 17-game campaign every season.[2] However, each game or battle is fought against one team at a time. Two or three mediocre teams cannot form an alliance and combine their forces to defeat a first-place team. There are no alliances or coalitions as football leagues reside in a Hobbesian world where everyone is out on their own.

Alliances are as old as warfare itself. Thucydides, the world's first military historian described in detail the alliances of Athens and Sparta in his *History of the Peloponnesian War*. It was Sparta who discovered the value of forming an alliance with Persia, a power that possessed a maritime capability they lacked that proved decisive in their eventual victory. One of today's premier military historians, Sir Lawrence Freedman, emphasizes coalition formation as one of three common fundamental features throughout history. The other two being deception and the instrumental use of violence.[3]Post World War II, America's constellation of alliances such as NATO, and with various nations in the Pacific are a decisive national advantage.

Fighting as a coalition is the foundation upon how the U.S. military fights its wars. The key ingredient to victory in the American Revolution was the presence of the French Army led by General Rochambeau, and the French Navy led by Admiral DeGrasse. The value of allies proved its worth in later wars such as the First and Second World Wars, the Korean War, Desert Storm, and most recently in Afghanistan and Iraq.

In the modern era, alliances and partnerships remain as the U.S. military's strategic center of gravity.[4]Since World War I, the United States has fought as part of an alliance or coalition in every major war, which includes an expansion of permanent alliances following World War II.[5] When allies and partners contribute to a U.S. military operation, the U.S. gains a multitude of advantages including international legitimacy, expanded access, overflight, and overland permissions. The ability to project power overseas to ensure America is fighting an away game is dependent on our allies. Allies and partners contribute niche capabilities to operations where the U.S. may have a capability shortfall. For example, the Dutch are known for their air defense capabilities, while the Czech military has ample nuclear, biological, and chemical (NBC) defense capabilities. Further, allies and partners contribute their intelligence collection and analysis, broadening the perspective of military and civilian decision makers.

13.3 Rules of the Game

While there are rules that govern both football and war, there is no guarantee that adversaries and enemies will respect the international rules and norms of warfare. Indeed, America often fights wars and more broadly executes its foreign policy with an expectation that other nations will follow international norms. However, adversaries such as China are often quick to point out that they were not present at the table when the rules were written. Moreover, some nations and various non-state actors and terrorist organizations simply refuse to follow these rules.

On the football field, players and coaches play the game with the understanding that players from both sides will abide by the rules of the game. Indeed, football games are played with a stable of referees, linesmen, umpires, and other officials who are charged with enforcing the rules. Each team must gain 10 yards for a fresh set of downs, field goals earn three points and touchdowns are worth six points for either team. Both teams begin each half with three timeouts. Failure to follow agreed upon rules of the game lead to on the field penalties, or off the field suspensions and fines.

In war, the rules of the game are complicated. While nations tend to sign international agreements that govern the rules of armed conflict, there is no guarantee that nations will follow said rules. Perhaps one of the more absurd international agreements was the Kellogg-Briand Pact of 1928. This agreement, initially signed by fifteen nations (with an additional 47 later) and ratified by the U.S. Senate was an outgrowth of the experience of the First World War. The pact outlawed war as an instrument of national policy and required members to settle international disputes via peaceful means. The treaty soon failed with the Japanese invasion of Manchuria. Moreover, the rise of dictatorships and authoritarian governments during the great depression would lead to the Second World War, a conflict even deadlier than the first. As we learned from Clausewitz, war is a method to attain policy objectives, and attempts to eliminate war as a policy choice are fruitless endeavors.

There have been other treaties and agreements seeking to place limits on war, and the conduct in war. The Ottawa Treaty on Land Mines signed in 1997 seeks the permanent prohibition on the "use, stockpiling, production, and transfer of anti-personnel mines."[6] The Chemical Weapons Convention (CWC) banned the use of chemical weapons and required their destruction. This agreement added to the 1925 Geneva Protocol which outlawed the use but not the possession of chemical weapons.[7] More recently, international organizations are seeking to limit the use of artificial intelligence systems that decide to kill humans on the battlefield in the hopes of keeping a "human in the loop." The American way of war leans towards following international agreements without necessarily signing them.

Rules, just as tactics, operations, and strategy evolve over time. In football, the evolution of the rules tends to favor a more offensive game that fans of the game tend to find more appealing. One of the earliest changes to the rules was the permission for an offense to attempt the forward pass. Teams that took advantage of this rule change scored major upsets over heavily favored opponents. An example of this was the famous Notre Dame 35—13 victory over Army in 1913 featuring Knute Rockne, which was enabled by the first extensive use of the forward pass.

The NFL rulebook is 94 pages long and only covers the NFL. Each league from youth to high school through college has its own rulebooks with enforceable violations. More recently, rules change in football include limiting the contact between a defensive back and a receiver and limiting the areas where a defensive player may contact the quarterback. The NFL for example has a financial interest in protecting its star players and quarterbacks, thus ensuring the most competitive play on the field. Following injuries to Tom Brady in 2008, the NFL enacted the "Brady Rule" that barred defensive players from lunging toward the knees of quarterbacks.[8] More recently, the NFL changed the rules regulating and limiting how and where a defensive player can contact offensive receivers. Each of these rule changes enhances the offensive flow, reinforcing the entertainment and for-profit objectives of the game.

Each level of the game has a slate of referees, umpires, and judges to oversee the game and enforce the rules. The number of officials at each level varies, with the highest level of college and the NFL having a full slate while youth football games may have as little as 2 or 3 referees. Penalties witnessed by the referees result in a yellow flag tossed to the ground, followed by a pre-determined punishment based on the foul. Five yards for offsides, ten yards for offensive holding, and fifteen yards for unsportsmanlike conduct. The referees keep the time, rule if a pass is caught in the air or incomplete, and judge which team's player recovered a fumble at the bottom of a scrum. This enforcement of the rules is opposite the American way of war which relies on self-reporting and enforcement of the rules of warfare. America prefers to prosecute its service members when they commit crimes in combat zones in lieu of an international organization. There is a history of prosecution for war crimes dating back to the Nuremberg Trials after World War Two and more recently the International Criminal Court's prosecution of war criminals from the Balkans. However, these prosecutions occur following each respective war, and not during the conflict.

13.4 Set Pieces

At the start of every play, 11 players line up on each side of the line of scrimmage. Both the offense and defense have players in specific positions such as quarterbacks, linemen, running backs, receivers, and tight-ends on offense, and linemen, linebackers, cornerbacks, and safeties on defense. Further, each team has a limit to the number of players that can be on the game day roster. In addition to the players, each team has a similar structure of coaches, including the head coach, offensive and defensive coordinators, and various position coaches. On the field, the same number of referees officiate the game week to week. While formations and plays may vary, one team will never gain a competitive advantage by putting more players between the sidelines than the opposition.

Before the modern era of war, armies would fight in two distinct ways. First, in a siege, where one army surrounded a town or a fortress, and slowly moved forward, simultaneously bombing, and starving the population. The second way was through battle on an open field. Armies lined up in set piece formations, exchanging musket and cannon fire. With advancements in technology, the size and lethality of the battlefield expanded culminating in global conflicts. Today it is a thin line that separates local brush fires from regional and global conflagrations.

The predictable nature of a team's structure is the opposite of what makes up the modern battlefield and contemporary operating environment. In addition to uniformed service members, the battlefield will have peacekeepers from the United Nations, contractors, members of the press, and non-governmental organizations such as the International Red Cross (IRC), Doctors Without Borders, and UNICEF. State and non-state actors alike will appear on the battlefield with or without an invitation. These organizations arrive in contingencies spanning from major combat operations to humanitarian assistance missions following a natural disaster. Each of these actors and organizations has its own internal objectives and purpose, and often these objectives are in competition with others.

13.5 Predictable Schedules

Football teams know when and where they will play. Before the start of every season, team schedules at all levels of football are released and teams know the time and location of every regular season game, and they know who will be standing on the opposing sideline. Although playoff schedules may vary, teams will know who the opponent will be a week in advance as they make their way through the post-season bracket. The predictable schedule enables teams to plan out their practices, individual conditioning, and to study relevant film to ensure the best performance on game day. There is no surprise game, and teams can't just show up to another team's stadium, line up on the field, and demand a contest.

Militaries rarely have the luxury of knowing who, when, or where they will fight. Although the combatant commands such as Central Command, European Command, and Africa Command keep a set of war plans on the shelf, those plans may never come into fruition. Indeed, there was zero planning for the U.S. military Prior to the attacks of September 11th, there was zero planning for the United States to invade Afghanistan as the nation's military and political leadership was focused on other enemies and adversaries. Moreover, militaries are often asked to conduct a wide variety of non-combat operations such as delivering humanitarian assistance following a natural disaster, which often comes as a surprise.

More than wars being unpredictable, battles and tactical engagements are often unexpected, at least to one side. War is replete with examples of adversarial forces crossing paths while moving from one point to another. The idea of movement to contact is a method to move a formation with the hope of meeting the enemy at an unexpected time and place. Unexpected contact can give one side the advantage of initiative and surprise.

The U.S. military does conduct detailed planning for various scenarios while developing capabilities to execute these on the shelf war plans, but the timing of when these plans will go into effect is uncertain and may not ever occur at all. Despite decades of planning and preparation, the U.S. military never had to fight the Russian hordes coming across the Fulda Gap.

Of course, some conflicts are predictable. During the interwar years between WWI and WWII, the U.S. Army and U.S. Navy spent time preparing plans to mobilize and fight the Japanese in the Pacific and the Germans in Europe. Various "color" plans, such as War Plan Orange (Japan) and War Plan Black (Germany), provided a basis for the Rainbow Plans—so-called because they envisioned war against multiple opponents (colors) simultaneously. Rainbow Five was the starting point for the United States entering the Second World War.[9] The 2003 invasion of Iraq was planned for a long time and in a sense a sequel to the 1991 Gulf War.

13.6 Homefield Advantage

In football, as in most team sports, there is an advantage to playing at home in front of a friendly crowd. While the home field advantage does not determine the winner and loser, it often serves as a competitive advantage that can sway the outcome of a game between two relatively evenly matched teams. Often, towards the end of the regular season football teams who clinched a playoff birth will continue to field a normal starting lineup with the hope of securing the home field advantage in the postseason.

The American way of war, at least since the end of the Civil War in 1865 tends to occur in foreign countries. Indeed, according to the historian Williamson Murray, the American way of war centers on power-projection, or the ability to deploy military forces overseas.[10] The Spanish – American War, World Wars I and II, Korea, Vietnam, Iraq, and Afghanistan were all away games for the United States. However, the advantage of projecting power overseas and fighting abroad changes when fighting an insurgency or involving the military inside of another nation's civil war. The inability to gain strategic victory in Iraqi Freedom and Enduring Freedom following initial conventional military success serve as home field advantage examples.

13.7 The Ending

Every football game ends with a winner, a loser, or in a tie. After four quarters and at times overtime, the game comes to an end. Moreover, each team will complete their games at the end of the regular season, with a loss in the playoffs, or as the league champion. There are defined winners and losers at each level of the game. While the competition continues year in and year out, the timeframe of each tactical engagement on the field is pre-determined.

War tends to be a bit fuzzier. There are instances of clear-cut winners and losers as in the case of the United States and the allies prevailing over the Axis powers in the Second World War. Other wars tend to continue without a clear-cut winner and loser, but rather agreements to suspend or end the fighting. Indeed, the border between North and South Korea serves as a constant reminder that the two sides agreed to an armistice and are still in a state of war.

13.8 Divergent Purposes

It is worth repeating the divergent purposes of football and war to conclude this chapter. At the highest levels, football is a business. The overarching goal of an owner of a NFL team is to earn revenue. Fielding a successful team is how this occurs. Indeed, the support of the local population is paramount to the financial stability of a franchise. Generating more wealth in ticket sales and television contracts will enhance a team's ability to acquire the best talent. Winning at the strategic level means maintaining a period of sustained excellence (multiple divisions or conference titles, multiple Super Bowl championships, etc. …) over multiple years.

The decisions to go to war and how to fight a war is a serious business for serious people. War is not a business, and the goal of senior military leaders is not to make money for their service or the Department of Defense but rather to fight and win the nation's wars. Nations fight wars to attain a policy outcome in support of their national interests. Fighting wars costs money and fighting wars costs lives. Getting the decisions that govern war wrong are catastrophic for the people that fight them, and for the nation that sends them.

Notes

1. Mark Edmundson, *Why Football Matters: My Education in the Game.* (New York, NY: Penguin Books, 2014).
2. The 17-game season began in 2021. It was previously set at 16 games (not including the payoffs).
3. Lawrence Freedman, *Strategy: A History.* (New York, NY: Oxford University Press, 2013), 3.
4. Alliances are permanent with formal defense agreements while partnerships are temporary and often geared towards specific objectives in space and time.
5. Joseph F. Dunford, "Allies and Partners are our Strategic Center of Gravity. *Joint Forces Quarterly*, 87, 4th Quarter 2017), 4–5.
6. *Convention On The Prohibition Of The Use, Stockpiling, Production And Transfer Of Anti-Personnel Mines And On Their Destruction.* Retrieved from: https://geneva-s3.unoda.org/static-unoda-site/pages/templates/anti-personnel-landmines-convention/APLC%2BEnglish.pdf
7. United Nations Office for Disarmament Affairs: https://www.un.org/disarmament/wmd/chemical/
8. *2020 Official Playing Rules of the National Football League.* Pg. 51. Retrieved from: https://operations.nfl.com/media/4693/2020-nfl-rulebook.pdf
9. War Plan Orange and War Plan Black were "color" plans crafted for war with specific adversaries. They usually assumed the US would fight alone. The Rainbow Plans were a separate series of plans for global war developed in the run-up to World War II. They were designated "rainbow" because they anticipated a war with multiple colors—allies and opponents—fighting simultaneously.
10. Williamson Murray, *America and the Future of War: The Past as a Prologue.* (Stanford, CA: Hoover Institution Press, 2017), Kindle.

Chapter 14
Conclusion / Epilogue

So there it is, a wild ride using football to explain military concepts from the armchair quarterback's perspective, and using military concepts to explain football from an armchair general's perspective.

My last game in a football uniform I broke down in tears on the sideline. It was the 4th quarter of the Nassau County championships, and we were down 31–0 to Lawrence High School. This was the end of the line, ten years of work and effort, ten years of suiting up for practice and sweating thought hot summer and early fall practices would end in abysmal failure. I had tears running down my cheeks, not because of the score, or the way I missed a block or an assignment during the game, but because I understood a portion of my life was over. The friends I made on the football field over the previous decade would begin to move in different directions. Today we make snarky comments to each other on social media, but somehow only see one another in person at reunions, weddings, and funerals. For a while the funerals were for parents who sat in the bleachers in all sorts of weather going unappreciated by 10-year-old kids whose universe was centered on ourselves. In a few years the funerals won't be for our parents, they will be for each other. Such is the reality of adulthood.

Like I said, in life we tend to fail more than we succeed. But that last game signified more than the end of the season, it signified the end of childhood. My focus would turn to preparing college applications and thinking about what I was going to do with the rest of my life. I had already enlisted in the Army Reserves and figured a life in uniform is what lay ahead.

While in uniform, some elements of football continue. At every duty station in America and abroad, soldiers, sailors, airman, and marines often occupy football fields for morale, welfare, and recreation (MWR) sponsored flag-football leagues. Although there is no tackling, it doesn't necessarily mean there is no contact, and games between rival units can bring out a long-repressed competitiveness in everyone participating. In a way it's an attempt to keep a grip on what remains of our youth.

D. Sukman, *American Football and the American Way of War*,
https://doi.org/10.1007/978-3-031-55345-5_14

It's hard to move on from anything that you spend a significant portion of your life doing. Should I live to 100, 10% of my life would involve donning a football uniform and competing on the gridiron. When I eventually leave the military, over 25% of my life will be in the army uniform. Leaving the service will be an emotional event, but I can look back to my days of youth to understand how life's transitions work. Everything begins and everything ends.

Annex A
Quarterbacks and Generals

War is too important to be left to the generals – Georges Clemenceau.

Find a mistake-proof quarterback and you have this game won. – Woody Hayes.

The foundation of any military unit is leadership. From the tactical level of war through the strategic level of war, the demand for effective leadership is constant. Military forces equipped with the latest technology and trained to the highest standard can fail spectacularly when led by incompetent and toxic leaders. The same concept applies to football, where teams that collect the most individual talent can consistently lose through poor leadership. It is leadership that sets the pace for an organization's culture.

Military organizations often develop their own unique culture. This is true for each of the services (Army, Navy, Air Force, Marines, Space Force), as well as sub-units within each service. For example, the 82nd Airborne Division's culture, composed of paratroopers is quite distinct from the culture of the 3rd Infantry Division. In the Air Force, fighter pilots have a distinct culture from heavy pilots. And in the Navy, Surface Warfare Officers (Ship Drivers) had a culture set apart from the Submarine community.

Just as military forces develop their own cultures, so do football teams and their respective athletes. From high school, through college, through the ranks of the NFL, each team develops its own unique culture. Further, just as each military service has sub-cultures, football teams do as well. A team's offense and defense may have their own culture and cliques, as will various positions such as lineman and running backs. The culture of a football team, much like a military organization's culture begins with the leadership.

Militaries and football organizations have similarities in the selection of their leaders. For the most part, military officers and football coaches rise through the ranks, with their elevation to the next grade correlating directly to their performance in previous assignments. Military officers rise to strategic leadership based on tactical level accomplishment, and head coaches in the NFL rise from the college level, and often spend significant time as assistant coaches and as offensive and defensive

D. Sukman, *American Football and the American Way of War*,
https://doi.org/10.1007/978-3-031-55345-5

coordinators on a professional team. However, tactical expertise does not equate to strategic acumen. Military history is filled with generals and admirals who excelled at tactics and failed when put into strategic leadership positions.

Character is at the forefront of both military and football leadership. Indeed, coaches' and commanders' character and reputation take decades to build, and an instant to destroy. Even the most accomplished leaders can soil their reputations through unethical and immoral behavior. For all his military innovations and accomplishments, General David Petraeus placed an asterisk next to his name when he passed classified information to his biographer with whom he was having an extramarital affair. Similarly, coaches such as Sean Peyton suffer damage to their reputation as occurred in the famous "bounty gate." When military leaders, famous football coaches, and athletes fail, it is front-page news. Americans place an enormous amount of trust in the military and hold the idea of a head coach in high regard.

If the action on the field is at the tactical level of football, then the apt comparison of the quarterback is to a commanding general. When two opposing football teams play each other, the buildup to the game includes framing the contest as one quarterback against the other. Staubach versus Bradshaw, Montana versus Elway, and Manning versus Brady are just a few of the famous quarterback matchups we have witnessed over the past few decades. Generals, like quarterbacks, are the face of their organization and represent the success or failure of their unit or team. For Example, the American Civil War is often spoken in terms of Grant versus Lee as much as it is the North versus the South. There are set categories that we can bin each of them into.

The Tom Brady GOAT Division. After winning seven Super Bowls, appearing in ten, winning more than a dozen division titles, and three Most Valuable Player (MVP) awards, Tom Brady is generally regarded as the greatest quarterback of all time. There are a handful of American generals that fall into this category. This includes Generals George Washington, U.S. Grant, William T. Sherman, and George C. Marshall. These are the generals that typically garner the credit for leading the United States to victory in the decisive wars of our nation.

The Peyton Manning Division. This is the category of hall of fame caliber generals who fall just short of falling into the greatest of all-time arguments. Peyton Manning won two Super Bowls, set multiple passing records (yards, touchdowns, etc..) throughout his career, and is considered one of, but not the all-time great. Generals in the Peyton Manning Division include Nathaniel Greene, Phillip Sheridan, and George Patton. Each of these officers had distinguished careers and excelled at war but cannot match the resume of the GOAT division.

The Dan Marino Division. Dan Marino was the most prolific passer of his time. Marino put up passing statistics that nobody would match until the character and rules of football changed. Marino won a ton of games in the regular season, but alas, never won a Super Bowl. The military equivalent to this division is generals who win tactical level battles but fail to turn their tactical success into operational or strategic victory. William Westmoreland's forces won multiple tactical engagements throughout the war in Vietnam, only to have history judge him as a failure for not winning the war. Confederate generals such as Robert E. Lee, who demonstrated tactical brilliance in battle such as Chancellorsville, but ultimately could not develop a winning strategy fit this mold.[1] The goal for generals is to win wars, and the goal for quarterbacks is to win championships.

The Joe Flacco Division. This division is reserved for generals who, like Joe Flacco can't decide if they are elite, run of the mill, or not so good. Flacco had a fantastic playoff run back in 2012 but mostly performed at an average to above average level throughout his career.[2] It is a clear case of regression to the mean, yet the debate continues. Generals such as Mark W. Clark fall into this division. Clark has a distinguished career, at one point becoming the youngest 4-Star in the U.S. Army. He had his great victory in Italy (despite Monte Casino) liberating Rome on the same day of the Normandy landings. However, his victory in Italy is marked by the escape of the German 10th Army. Where Mark Clark likes among the great leaders of the Second World War is always a great debate.

The Trent Dilfer Division. In 2000, Trent Dilfer won a Super Bowl with the Baltimore Ravens. In 2001, Dilfer was playing for the Seattle Seahawks. Deservedly or not, Dilfer earned a reputation as a game manager. Game managers are quarterbacks who succeed thanks to other aspects of the team. A strong defense or rushing attack enables game managers to rack up victories despite having average talent. There are general officers and senior leaders who succeed in large thanks to the supporting team around them. They may have a strong staff, a terrific XO and deputy commander, or phenomenal NCOs that make up for their many faults.

The Warren Moon/Joe Montana Division. Both quarterbacks had hall of fame careers and deserve recognition as two of the best that ever played the game. They put up statistics unmatched in their respective eras. However, critical to their success was the offensive system they led. Warren Moon played in the Run and Shoot Offense, known for its prolific passing attack. Montana played in Bill Walsh's West Coast Offense scheme, which revolutionized the game. Arguably, both Moon and Montana would find success in any system, but they played in the systems most beneficial to their talents. Moreover, after Montana left San Francisco, his former backup quarterback Steve Young would win a championship in the same system (albeit under a different head coach). In some ways all quarterbacks prosper when put in the right system with the right coach, and getting talent to blossom relies on the right coach create the right conditions. Over the years, our military produced numerous systems generals who could succeed in one type of warfare but could not prove their worth in others. Some do very well in maneuver warfare but fail in a counterinsurgency environment. The talents required in one do not always apply to another. Further, as General Petraeus discovered, what works in Iraq does not translate to success in Afghanistan.

The Alex Smith Division. Some quarterbacks succeed by the design of the system they play in. Throughout his career, Alex Smith has succeeded due to coaches designing offenses that fit his skill set. Scharnhorst, the creator of the Prussian Staff designed a system that did not rely on individual talent. Indeed, the lesson Scharnhorst learned from the Napoleonic Wars was that Prussia could not afford to wait for a version of Napoleon to rise in the ranks. Scharnhorst then created the Prussian General Staff system along with the famous Kriegsacademie. In the modern era, the

United States follows this mold, creating dozens of GOFOs every year through our assignment and professional education system. Everyone who rises to the top of the ranks can be replaced by the next cohort of officers.

The Ryan Leaf Division. Ryan Leaf was a star in college. Looking back, it's comical that in 1998 people debated whether he should be drafted ahead of Peyton Manning. Ryan Leaf was unable to handle the pressure and commitments of life in the National Football League. In short order, Leaf would find himself out of football and serving time for numerous drug-related offenses. Just like in professional sports, there are men and women selected for battalion and brigade command or promoted to general officer and admiral but were clearly not ready. Some not only should have never been promoted to flag rank, but rather they belong in jail for their respective ethical and moral lapses. Former brigadier general (now retired lieutenant colonel) Jeffrey Sinclair falls into this category.[3] One can find dozens of other examples that often appear on the front page of Army, Navy, and Air Force Times.

The Frank Reich/Nick Foles Division. In the early 1990s, Frank Reich was a backup to Jim Kelly on the Buffalo Bills. In the 1993 playoffs, Reich led the Bills to a 32-point comeback against the Houston Oilers, the largest comeback in NFL history. More recently, in 2017, Nick Foles took over the play calling from Carsen Weintz to lead the Philadelphia Eagles to a Super Bowl victory. Back-up quarterbacks rarely have the skill set of the starting quarterback but must be prepared to step in at any time. Not every general officer gets the chance to command. Many spend their time as J-Code or G-Code directors and assistant directors. However, their work in these key positions is paramount to the success of their respective organization. When called upon to lead in the commander's absence, they step up to the plate. Think about Brig. Gen. McAuliffe in his role as assistant division commander of the 101st in World War II is an example of a GOFO in this division.

The Ryan Fitzpatrick Division. Throughout his career, Ryan Fitzpatrick has served as the starting quarterback for eight teams.[4] For each team, Fitzpatrick has put up solid statistics but is often seen as a transitory, rather than a long-term franchise quarterback.[5] The military equivalent to Ryan Fitzpatrick is the generals who show up to an organization with the intent to make some serious and much needed change. Everyone in the command is excited after some early success but gives up hope after the general leaves for a more senior position after only nine months, only to offer the next command more false hope. In the military, continuity matters. Subordinates deserve dedicated time from senior leaders to develop and implement change. Merely placing a leader into an organization for a block check does more harm than good to the force at large.

The Gary Kubiak Division. Gary Kubiak was a long-term back-up under John Elway in the Denver Broncos organization. He never achieved fame as a quarterback at the tactical level but used his time as a backup to enhance his thinking, coaching, and leadership abilities. Later in life, Kubiak would win a Super Bowl as head coach of the Denver Broncos. This shows that some people perform better at the operational and strategic level rather than the tactical level. Colin Powell never commanded a division, and Eisenhower did not command above the battalion level. Once again, tactical success does not always translate to strategic brilliance, while some of the best operational and strategic leaders never stood out as exceptional at the tactical level.

The Vince Young Division. Vince Young earned the Heisman and won a national championship while playing for Texas. However, his professional success would never match what he achieved in college (Ditto for his Rose Bowl opponent Matt Leinart). These are the generals that succeeded at the lower ranks, and enjoyed some success as junior general officers, but failed spectacularly at the highest level of command. Think General Hooker at Chancellorsville or General William Westmoreland in Vietnam. With regards to Westmoreland, he played a critical role in the Battle of Kassarine Pass, often seen as the turning point in the American North Africa Campaign. However, his success as a tactical battalion commander did not translate to victory as the commanding general of American Forces in Vietnam.

Annex B
George Carlin's Comparison of Football to Baseball

This monologue ranks high in George Carlin's lifelong comedy career in terms of best routines. George Carlin used a version of this as his opening monologue as the guest host of the very first episode of *Saturday Night Live* in October of 1975. The monologue below appears in his book *Brain Droppings*.[6] When reading, it is best to read it with the George Carlin voice in your head.

> Baseball is different from any other sport, very different. For instance, in most sports you score points or goals; in baseball you score runs. In most sports the ball, or object, is put in play by the offensive team; in baseball the defensive team puts the ball in play, and only the defense is allowed to touch the ball. In fact, in baseball if an offensive player touches the ball intentionally, he's out; sometimes unintentionally, he's out.
>
> Also: in football, basketball, soccer, volleyball, and all sports played with a ball, you score with the ball and in baseball the ball prevents you from scoring.
>
> In most sports the team is run by a coach; in baseball the team is run by a manager. And only in baseball does the manager or coach wear the same clothing the players do. If you'd ever seen John Madden in his Oakland Raiders uniform, you'd know the reason for this custom.
>
> Now, I've mentioned football. Baseball & football are the two most popular spectator sports in this country. And as such, it seems they ought to be able to tell us something about ourselves and our values.
>
> I enjoy comparing baseball and football:
>
> Baseball is a nineteenth-century pastoral game.
> Football is a twentieth-century technological struggle.
>
> Baseball is played on a diamond, in a park. The baseball park!
> Football is played on a gridiron, in a stadium, sometimes called Soldier Field or War Memorial Stadium.
>
> Baseball begins in the spring, the season of new life.
> Football begins in the fall, when everything's dying.

D. Sukman, *American Football and the American Way of War*,
https://doi.org/10.1007/978-3-031-55345-5

In football you wear a helmet.

In baseball you wear a cap.

Football is concerned with downs - what down is it?

Baseball is concerned with ups - who's up?

In football you receive a penalty.

In baseball you make an error.

In football the specialist comes in to kick.

In baseball the specialist comes in to relieve somebody.

Football has hitting, clipping, spearing, piling on, personal fouls, late hitting and unnecessary roughness.

Baseball has the sacrifice.

Football is played in any kind of weather: rain, snow, sleet, hail, fog…

In baseball, if it rains, we don't go out to play.

Baseball has the seventh inning stretch.

Football has the two-minute warning.

Baseball has no time limit: we don't know when it's gonna end - might have extra innings.

Football is rigidly timed, and it will end even if we've got to go to sudden death.

In baseball, during the game, in the stands, there's kind of a picnic feeling; emotions may run high or low, but there's not too much unpleasantness.

In football, during the game in the stands, you can be sure that at least twenty-seven times you're capable of taking the life of a fellow human being.

And finally, the objectives of the two games are completely different:

In football the object is for the quarterback, also known as the field general, to be on target with his aerial assault, riddling the defense by hitting his receivers with deadly accuracy in spite of the blitz, even if he has to use shotgun. With short bullet passes and long bombs, he marches his troops into enemy territory, balancing this aerial assault with a sustained ground attack that punches holes in the forward wall of the enemy's defensive line.

In baseball the object is to go home! And to be safe! - I hope I'll be safe at home!

Rest in Peace George.

Notes

1. This is not meant to compare the character of Dan Marino to Robert E. Lee. By all accounts, Dan Marino is an honorable person, unlike Bobby Lee who betrayed his country and fought to maintain the institution of slavery.
2. Average for an NFL quarterback, which is exceptional compared to 99.99% of the world's population.

3. Former BG Sinclair, now retired LTC Sinclair admitted to multiple improper relationships.

   Kelly Twedell, "Army General gets fine, no jail in sex case." Reuters. 20 March 2014. Retrieved from: https://www.reuters.com/article/us-usa-courtmartial-sinclair-idUSBREA2J0XV20140321/.
4. St. Louis Rams, Cincinnati Bengals, Buffalo Bills, Tennessee Titans, Houston Texans, New York Jets, Tampa Bay Buccaneers, and Miami Dolphins.
5. This pattern continued in 2020. Ryan Fitzpatrick began the season as a starter for the Miami Dolphins only to be replaced by Tua Tagovailoa.
6. George Carlin, *Brain Droppings*. (New York, NY: Hyperion, 1997), 52–53.

Bibliography

Books

1. Lars Anderson, *Carlisle Vs. Army: Jim Thorpe, Dwight Eisenhower, Pop Warner, and the Forgotten Story of Football's Greatest Battle.* (New York: Random House, 2008).
2. Lars Anderson, *The All Americans.* (New York, NY: St. Martin's Press, 2004).
3. Beth L. Bailey. *America's Army: Making the All-Volunteer Force* (Cambridge, MA: Bellcamp Press, 2009).
4. Beth Bailey, *An Army Afire: How the US Army Confronted Its Racial Crisis in the Vietnam Era.* (Chapel Hill: University of North Carolina Press, 2023, 2023)
5. William J. Baker and John M. Carroll, ed. *Sports in Modern America.* (St. Louis, MO: River City Publishers, 1983)
6. Steve Belichick, *Football Scouting Methods.* (Mansfield, CT: Martino Publishing, 2011).
7. Jesse Berret, *Pigskin Nation: How the NFL Remade American Politics.* (University of Illinois Press: Springfield, I, 2018).
8. Richard K. Betts, ed. In *Conflict After the Cold War: Arguments on Causes of War and Peace,* 4th ed.(New York: Routledge, 2016).
9. Timothy P. Brown, *Fields of Friendly Strife: The Doughboys and Sailors of the WWI Rose Bowls.* (West Bloomfield, MI: Brown House Publishing, 2017).
10. Timothy P. Brown, *How Football Became Football: 150 Years of the Game's Evolution.* (West Bloomfield, MI: Brown House Publishing, 2020).
11. George Carlin, *Brain Droppings.* (New York, NY: Hyperion, 1997).
12. Will Carroll, Tyler Brooke, and Peter King, *The Science of Football: The Math, Technology, and Data Behind America's Game.* (New York, NY: Sports Publishing, 2022).
13. Carl von Clausewitz, On War, eds. Michael Howard and Peter Paret (Princeton, NJ: Princeton University Press, 1989).
14. Edward M. Coffman, *The Regulars: The American Army, 1898-1941.* (Cambridge, MA: Belknap Press, 2007).
15. Eliot A. Cohen, *Supreme Command: Soldiers, Statesmen, and Leadership in Wartime* (New York, Anchor Books, 2002).
16. Robert Coram, *Boyd: The Fighter Pilot Who Changed the Art of War.* (New York, Hachette Book Group, 2002).
17. Russ Crawford, *Le Football: A History of American Football in France.* (Lincoln, NE: University of Nebraska Press, 2016).
18. Department of Defense, *2022 National Defense Strategy.*
19. Carlo D'Este, *Eisenhower: A Soldier's Life,* (New York, NY: Henry Holt, 2002).
20. Giulio Douhet, *Command of the Air.* (Pickle Partner Publishing, 1942).

D. Sukman, *American Football and the American Way of War*,
https://doi.org/10.1007/978-3-031-55345-5

21. Whitfield B. East, A Historical Review and Analysis of Army Physical Readiness Training and Assessment. (Fort Leavenworth, KS: Combat Studies Institute Press, 2013).
22. Mark Edmundson, *Why Football Matters: My Education in the Game.* (New York, NY: Penguin Books, 2014).
23. John Eisenberg, *The League: How Five Rivals Created the NFL and Launched a Sports Empire.* (New York: Basic Books, 2018).
24. Mark Fainaru-Wada and Steve Fainaru, *League of Denial: The NFL, Concussions, and the Battle for Truth*, (New York, Three Rivers Press, 2013).
25. Doug Farrar. *The Genius of Desperation. The Schematic Innovations That Made the Modern NFL.* (Chicago, IL: Triumph Books, 2018).
26. John Feinstein, *A Civil War: Army vs. Navy - A Year Inside College Football's Purest Rivalry* (New York, NY: Back Bay Books, 1997).
27. John Feinstein, *Next Man Up: A Year Behind the Lines in Today's NFL.* (New York: Little Brown and Company, 2006).
28. Marvin E. Fletcher, *America's First Black General: Benjamin O. Davis Sr., 1880–1970.* (Lawrence, KS: University Press of Kansas, 1989).
29. Lawrence Freedman, *Deterrence.* (Malden, MA: Polity Press, 2004).
30. Lawrence Freedman, *Strategy: A History.* (New York, NY: Oxford University Press, 2013).
31. Francis Fukuyama, *The End of History, and the Last Man.* (New York, NY: Simon & Schuster, 1992).
32. John P. Kotter, *Leading Change.* (Boston, MA: Harvard Business School Press, 2012).
33. John. Lewis Gaddis. *On Grand Strategy.* (New York, NY: Random House, 2018).
34. Bob Glauber, *Guts and Genius: The Story of Three Unlikely Coaches Who Came to Dominate the NFL in the '80s*, (New York, NY: Grand Central Publishing, 2018).
35. Mike Guardia, *Crusader: General Donn Starry and the Army of His Times* (Philadelphia, PA: Casemate Publishers 2018).
36. S.C. Gwynne, *The Perfect Pass, Reinvention of Football.* (New York, NY: Simon and Schuster, 2016).
37. Nikolas Gvosdev, Jessica Blankshaim, and David Cooper, *Decision-Making in American Foreign Policy*, Cambridge, 2019).
38. David Halberstam, *The Education of a Coach.* (New York, NY: Hachette Books, 2012).
39. Dale R. Herspring, *The Pentagon and the Presidency: Civil-Military Relations from FDR to George W. Bush.* (Lawrence, KS: University Press of Kansas, 2005).
40. B.H. Liddell Hart. *Strategy.* Second Revised Edition. (New York, NY: First Meridian Printing, 1991).
41. Michael Holly. *War Room: The Legacy of Bill Belichick and the Art of Building the Perfect Team.* (New York: Harper Collins Publishers, 2011).
42. Trent Hone. *Learning War: The Evolution of Fighting Doctrine in the U.S. Navy; 1898–1945.* (Annapolis, MD: Naval Institute Press, 2018).
43. Michael Howard, *The Causes of War and Other Essays.* (Cambridge, MA: Harvard University Press), 1983.
44. Brian M. Ingrassia, *The Rise of Gridiron University: Higher Education's Uneasy Alliance with Big-Time Football.* (Lawrence, KS: University of Kansas Press, 2012).
45. Georgii Samoilovich Isserson, *The Evolution of the Operational Art.* (Fort Leavenworth, KS: Combat Studies Institute Press, 2013).
46. Ron Jaworski, *The Games that Changed the Game.* (New York, NY: ESPN books, 2010).
47. Benjamin M. Jensen. *Forging the Sword: Doctrinal Change in the U.S. Army,* (Stanford, CA: Stanford University Press, 2016).
48. David E. Johnson, *Fast Tanks and Heavy Bombers: Innovation in the U.S. Army, 1917–1945.* (Ithaca, NY: Cornell University Press, 2013).
49. James W. Johnson, *The Wow Boys: A Coach, A Team, and a Turning Point in College Football.* (Lincoln, NE: University of Nebraska Press, 2006).
50. *Joint Publication 1–02 Department of Defense Dictionary of Military and Associated Terms.* February 2016. (Washington, DC: The Joint Staff).

51. *Joint Publication 5–0: Joint Planning*. 10 December 2020. (Washington, DC: The Joint Staff).
52. Antoine-Henri Jomini, *The Art of War*. (Project Gutenberg Ebook edition, 28 September 2004).
53. Ian S. Kahnanowitz, *NFL 1970: The Inaugural Season of the New NFL*. (Mechanicsburg, PA: Sunbury Press, 2021).
54. Michael Keane, *Dictionary of Modern Strategy and Tactics*. (Annapolis, MD: Naval Institute Press. 2005).
55. Yvan J. Kelly, David Berri, and Victor A. Matheson, *The Economics of the Super Bowl: Players, Performers, and Cities*. (Cham, Switzerland: Palgrave Macmillan).
56. Brian Kilmeade, *The Games Do Count: America's Best and Brightest on the Power of Sports*. (New York, NY: Harper Collins Publishers, 2004).
57. Tim Layden, *Blood, Sweat, & Chalk: How the Geniuses of Football Created America's Favorite Game*. (New York, NY: Sports Illustrated Books, 2010).
58. Adam Lazarus, *Super Bowl Monday: From the Persian Gulf to the Shores of West Florida – The New York Giants, the Buffalo Bills, and Super Bowl XXV*. (New York: Taylor Trade Publishing, 2011).
59. Robert R. Leonhard. *Fighting By Minutes: Time and the Art of War*. (KC Publishing. Kansas City, MO, 2018).
60. Michael Lewis, *The Blind Side: Evolution of a Game*. (W. W. Norton & Company. New York, NY, 2007).
61. Michael Lombardi. *Gridiron Genius: A Master Class in Winning Championships and Building Dynasties in the NFL*. (New York, Radom House LLC, 2018).
62. Catherine Lutz, *Homefront: A Military City and the American 20th Century*, (Boston, MA: Beacon Press, 2001).
63. Edward N. Luttwak. *Strategy: The Logic of War and Peace*. Bostaon, MA: Harvard University Press 2002).
64. David Maraniss, *When Pride Still Mattered: A Life of Vince Lombardi*. (New York, NY: Simon and Schuster, 1999).
65. Jim Mattis. C*all Sign Chaos: Learning to Lead* (New York, NY: Random House Press, 2020).
66. David McCullough, *Truman*. (New York, NY: Simon and Schuster, 1992).
67. Donella H. Meadows, *Thinking in Systems*. (White River Junction, VT: Chelsea Green Publishing, 2008).
68. Doug Melville, *Invisible Generals: Rediscovering Family Legacy, and a Quest to Honor America's First Black Generals*. (New York, NY: Atria/Black Privilege Publishing, 2023).
69. John J. Miller, *The Big Scrum: How Teddy Roosevelt Saved Football*. (Harper Collins, 2011).
70. Williamson Murray, *American and the Future of War: The Past as a Prologue*. (Stanford, CA: Hoover Institution Press, 2017), Kindle.
71. Williamson Murray, *Military Adaptation in War: With Fear of Change*. (Cambridge: Cambridge University Press, 2011).
72. Cathal J. Nolan. *The Allure of Battle; A History of How Wars Have Been Won and Lost*. (Oxford University Press, 2017).
73. Michael Oriard, *Brand NFL: Making and Selling America's Favorite Sport*. (Chapel Hill: University of North Carolina Press, 2007).
74. Phillips Payson O'Brian, *How the War was Won: Air-Sea Power and Allied Victory in World War II*. (Cambridge, UK: Cambridge University Press, 2015).
75. William G. Pagonis and Jeffrey L. Cruikshank. *Moving Mountains: Lessons in Leadership and Logistics from the Gulf War.* (Boston: Harvard Business Review Press, 1992).
76. Jeff Pearlman, *Football for a Buck: The Crazy Rise and Crazier Demise of the USFL*, (New York: Mariner Books, 2019).
77. Andrew Roberts, *Masters and Commanders: How Four Titans Won the War in the West, 1941-1945*. (New York: Harper Publishing, 2009).
78. R.D. Rosen, *Tough Luck: Sid Luckman, Murder, Inc., and the Rise of the NFL*. (New York: Atlantic Monthly Press, 2019).
79. Thomas C. Schelling, *Arms and Influence*. (New Haven, CT: Yale University Press, 1966).

80. Stephen W. Sears, *Landscape Turned Red: The Battle of Antietam*. (New York: Mariner Books, 2003).
81. Chris Serb, War Football: *World War I and the Birth of the NFL*. (Lanham, MD: Rowman & Littlefield Inc, 2019).
82. Matthew Sherry, *Any Given Sunday, The NFL's Epic 100-year History in 20 Games*. (London, UK: Orion, 2020).
83. Robert B. Strassler, ed., *The Landmark Thucydides: A Comprehensive Guide to the Peloponnesian War* (Free Press, 1996).
84. Gordon R. Sullivan and Michael V. Harper. *Hope is Not a Method*. (New York: Broadway Books, 1997).
85. Amy Trask, *You Negotiate Like a Girl: Reflections on a Career in the National Football League*. (Chicago, IL: Triumph Books, 2016).
86. Sun Tzu, *The Art of War*. Trans. Samuel B. Griffith. (Oxford University, 1971).
87. Milan Vego, *Joint Operational Warfare: Theory and Practice*. (Newport, RI: U.S. Navy War College, 2007).
88. Sam Walker, *The Captain Class: A New Theory of Leadership*, (New York, Random House 2017).
89. Bill Walsh, *The Score Takes Care of Itself*. (New York: Penguin Group, 2009).
90. Kate Werren, *An American Uprising In Second World War England: Mutiny in the Duchy*. (Yorkshire: Pen and Sword Books, 2020).
91. Thomas Wildenberg, *All the Factors of Victory: Adm. Joseph Mason Reeves and the Origins of Carrier Airpower*. (Annapolis, MD: Naval Institute Press, 2003).
92. Bob Woodward, *Obama's Wars*. (New York: Simon and Schuster, 2010).
93. J.C. Wylie. *Military Strategy: A General Theory of Power Control*. (Annapolis, Md, Naval Institute Press. 1989).
94. David T. Zabecki, "Brigadier General Fox Conner" in *Pershing's Lieutenants: American Military Leadership in World War I*. ed. David Zabecki and Douglas Mastriano. (Osprey Publishing, 2020).
95. David T. Zabecki, *The General's War: Operational Level Command on the Western Front in 1918*. (Bloomington, IA: Indiana University Press, 2018).

Articles

96. Jonathan Alter, "Secrets From Inside the Obama War Room." *Newsweek* Web, 31 May 2010.
97. Gillian R. Brassil and Kevin Draper, "These Women Were N.F.L. "Firsts." They're Eager for Company. *New York Times*. 3 February 2021. Retrieved from: https://www.nytimes.com/2021/02/03/sports/football/nfl-women-coaches.html.
98. David F. Burrelli, "The Repeal of "Don't Ask, Don't Tell": Issues for Congress." Congressional Research Service. December 21, 2012. Retrieved from: https://fas.org/sgp/crs/misc/R42003.pdf.
99. Norm Carlson, "Gator Talk – Strange Tales of Gator Coaches." Retrieved from: https://floridagators.com/news/2010/11/14/19205.aspx.
100. Mike Cole, "How Brian Hoyer Played a Pivotal Role in Patriots Defense's Super Bowl Success. 11 February, 2019. Retrieved from: https://nesn.com/2019/02/how-brian-hoyer-played-pivotal-role-in-patriots-defenses-super-bowl-success/.
101. Philip A. Crowl, "Alfred Thayer Mahan: The Naval Historian." In *Makers of Modern Strategy: from Machiavelli to the Nuclear Age*, edited by Peter Paret. (Princeton, NJ: Princeton University Press, 1986),
102. Nick Cullather, "Bomb the Back to the Stone Age: An Etymology." History News Network. 2006. Retrieved from: https://historynewsnetwork.org/article/30347.

103. Dan De Luce, "Aircraft carrier failed to enforce social distancing, allowing Covid to spread: Pentagon watchdog." NBC News. 8 February 2021. Retrieved from: https://www.nbcnews.com/news/us-news/aircraft-carrier-failed-enforce-social-distancing-allowing-covid-spread-pentagon-n1257023.
104. Joseph F. Dunford, "Allies and Partners are our Strategic Center of Gravity. *Joint Forces Quarterly*, 87, 4th Quarter 2017).
105. Mark Edmundson, "Football is America's War Game." *Los Angeles Times*. 23 August, 2014. Retrieved from: https://www.latimes.com/opinion/op-ed/la-oe-edmundson-americas-and-war-connection-20140824-story.html.
106. Jeremy Fuchs, "What If ... Theodore Roosevelt's Son Hadn't Altered Football Forever?" *Sports Illustrated*. Retrieved from: https://www.si.com/specials/greatest-sports-what-ifs/college-football/2017/theodore-roosevelt-jr-harvard-football.
107. Andy Gallagher, "Who is Sarah Thomas? Super Bowl 55 female referee making history." 7 February 2021. Retrieved from: https://www.profootballnetwork.com/sarah-thomas-nfl-female-official/.
108. Mark Heisler, "Al Davis: the Machiavelli of Football." Los Angeles Times. 15 December, 1991. Retrieved from: https://www.latimes.com/sports/la-xpm-1991-12-15-la-sp-al-davis-heisler-story.html.
109. Chris Huston, "A Tribute to Heisman Veterans," (2019). Accessed 1 July 2020. Retrieved from: https://www.heisman.com/articles/a-tribute-to-heisman-veterans-2/.
110. Michael Janofsky, "Shell is First Black Coach in N.F. L. Since 20's." *The New York Times*. 4 October 1989. Retrieved from: https://www.nytimes.com/1989/10/04/sports/shell-is-first-black-coach-in-nfl-since-20-s.html.
111. Robert Jervis, "Cooperation Under the Security Dilemma," In *Conflict After the Cold War: Arguments on Causes of War and Peace*, 4th ed., edited by Richard K. Betts. (New York: Routledge, 2016).
112. Paul, Joan, Richard V. McGhee, and Helen Fant. "The Arrival and Ascendence of Black Athletes in the Southeastern Conference, 1966–1980." *Phylon (1960-)* 45, no. 4 (1984): 284–97. Accessed July 19, 2021. https://doi.org/10.2307/274909.
113. Johnny R Jones,. "Fundamental Truths of Air Power." In *William 'Billy" Michell's Air Power*. (Maxwell AFB, AL: Airpower Research Institute, September 1997).
114. Jill Lieber, "A New Blue Yonder: Former Air Force Pilot Chad Hennings Reports for Duty with the Cowboys." *Sports Illustrated*. 13 July 1992. Retrieved from: https://vault.si.com/vault/1992/07/13/a-new-blue-yonder-former-air-force-pilot-chad-hennings-reports-for-duty-with-the-cowboys.
115. Arthur Lykke. 1989. "Defining Military Strategy = E + W + M," *Military Review* 69, no. 5.
116. Michael Matheny. "The Fourth Level of Warfare." *Joint Forces Quarterly* 80, 1st Quarter 2016.
117. Kathleen J. McInnis, *Goldwater-Nichols at 30: Defense Reform and Issues for Congress* (Washington, DC: Congressional Research Service, May 11, 2016), 7–8.
118. Michael McKnight. "Second Nature: Why NFL Coaches are so Prone to Clock Management Blunders." *Sports Illustrated*. Issue 29 August–5 September 2016.
119. Jeffrey Meiser and Patrick Quirk, "Want Better Strategists? Start With a Better Definition of Strategy." *Real Clear Defense*. 9 July 2020. Retrieved from: https://www.academia.edu/43964729/Want_Better_Strategists_Start_With_a_Better_Definition_of_Strategy.
120. Thomas B. Morgan, "The American War Game." *Esquire Magazine*. October, 1965. 71–72, 142, 146–148. https://classic.esquire.com/issue/19651001/print.
121. Jim Morrison. 2010. "The Early History of Football's Forward Pass." *Smithsonian Magazine*. Accessed 23 October 2018. (https://www.smithsonianmag.com/history/the-early-history-of-footballs-forward-pass78015237/).
122. Steven Norris, "Remembering Georgia Tech's Heroes." 26 May, 2016. Retrieved from: https://www.news.gatech.edu/2016/05/26/remembering-georgia-techs-heroes.
123. Michael Oriard, "Flag Football: How the NFL became the American War Game." *Slate*. 17 November, 2009. Retrieved from: https://slate.com/culture/2009/11/how-the-nfl-became-the-american-war-game.html.

124. Oristano, Mark, "How Legendary NFL Coach Honed his Instincts in World War II." *HISTORYNET*. 13 February 2022. Retrieved from: https://www.historynet.com/how-legendary-nfl-coach-tom-landry-honed-his-fighting-instincts-in-world-war-ii/.
125. Kevin Patra, "NFL instituting changes to Rooney Rule." NFL.com 18 May 2020. Retrieved from: https://www.nfl.com/news/nfl-instituting-changes-to-rooney-rule.
126. Eyder Peralta, "Pentagon Paid Sports Teams Millions For "Paid Patriotism" Events." National Public Radio. 5 November, 2015. Retrieved from: https://www.npr.org/sections/thetwo-way/2015/11/05/454834662/pentagon-paid-sports-teams-millions-for-paid-patriotism-events.
127. Sarah Pruitt. "9 Things You May Not Know About Gerald Ford." 1 September 2018. Retrieved from: https://www.history.com/news/9-things-you-may-not-know-about-gerald-ford.
128. John Raughter, "College football great Lou Holtz reflects on being named Legion's 2019 " Good Guy"" The American Legion. (23 October, 2019). Retrieved from: https://www.legion.org/convention/246961/college-football-great-lou-holtz-reflects-being-named-legion%E2%80%99s-2019-%E2%80%98good-guy%E2%80%99.
129. Jack Snyder, "One World, Rival Theories," *Foreign Policy* no. 145 (Nov/Dec 2004).
130. John J.Tierney, "The Use Of Systems Theories In International Political Analysis." *World Affairs*, vol. 134, no. 4, 1972, pp. 306–324. *JSTOR*, www.jstor.org/stable/20671334. Accessed 7 June 2021.
131. Marybeth P. Ulrich, "A Primer on Civil-Military Relations for Senior Leaders." *In U.S. Army War College Guide to National Security Issues, Vol. II*, J. Boone Bartholomees Jr., ed., (Carlisle Barracks, PA: U.S. Army War College, June 2008).
132. Amir Vera, "Fritz Pollard: First Black Coach in the NFL." Retrieved from: https://www.azfamily.com/community/black_history_month/fritz-pollard-first-black-coach-in-the-nfl/article_dff033b2-33e2-5049-a3f2-1a83d1e5eb06.html.
133. David Vergun, "Sports Heroes Who Served: Tennessee Football Coach Served in World Wars I, II." DoD News. (17 November 2020) retrieved from: https://www.defense.gov/Explore/Features/Story/Article/2411661/sports-heroes-who-served-tennessee-football-coach-served-in-world-wars-i-ii/.
134. Kathryn Watson. "Fight against ISIS has shifted to "annihilation tactics," Mattis says." CBS News. 28 May 2018. Accessed 27 November 2018. Retrieved from: https://www.cbsnews.com/news/fight-against-isis-has-shifted-to-annihilation-tactics-mattis-says/.
135. John S. Watterson, "Political Football: Theodore Roosevelt, Woodrow Wilson, and the Gridiron Reform Movement." *Presidential Studies Quarterly*, 25.3 (1995).
136. Phillip Walter Wellman, "Afghanistan: Remembering the fallen of 2020." *Stars and Stripes*. 30 December, 2020. Retrieved from: https://www.stripes.com/news/middle-east/afghanistan-remembering-the-fallen-of-2020-1.656849.
137. James Weeks, "Football As A Metaphor For War." *American Heritage*. September/October 1988. Retrieved from: https://www.americanheritage.com/football-metaphor-war.
138. Paul Yingling. "A Failure of Generalship," Armed Forces Journal. 1 May 2007. Retrieved from: http://armedforcesjournal.com/a-failure-in-generalship/.

Web Sites

139. *2020 Official Playing Rules of the National Football League*. Pg. 51. Retrieved from: https://operations.nfl.com/media/4693/2020-nfl-rulebook.pdf.
140. National Football Foundation. Retrieved from: https://footballfoundation.org/sports/general/roster/gen--james-a---van-fleet/80.
141. Presidents Who Played Football." Retrieved from: http://www.profootballhof.com/presidents-who-played-football/.
142. "The Heisman Trophy Winners List" Accessed 1 July 2020. Retrieved from: https://www.heisman.com/heisman-winners/.

143. Pro Football Reference Website. Retrieved from: https://www.pro-football-reference.com/boxscores/199101270buf.htm.
144. *Convention On The Prohibition Of The Use, Stockpiling, Production And Transfer Of Anti-Personnel Mines And On Their Destruction.* Retrieved from: https://geneva-s3.unoda.org/static-unoda-site/pages/templates/anti-personnel-landmines-convention/APLC%2BEnglish.pdf.
145. United Nations Office for Disarmament Affairs: https://www.un.org/disarmament/wmd/chemical/.
146. Bill Parcells. "Parcellsisms: You are What Your Record says you are. Interview with Nunyo Demasio. 7 October 2016. Retrieved from: https://www.youtube.com/watch?v=a1dtAVNEnkQ.

Index

D. Sukman, *American Football and the American Way of War*,
https://doi.org/10.1007/978-3-031-55345-5

SPRINGER NATURE

GPSR Compliance

The European Union's (EU) General Product Safety Regulation (GPSR) is a set of rules that requires consumer products to be safe and our obligations to ensure this.

If you have any concerns about our products, you can contact us on ProductSafety@springernature.com

In case Publisher is established outside the EU, the EU authorized representative is:

Springer Nature Customer Service Center GmbH
Europaplatz 3
69115 Heidelberg, Germany

FSC
www.fsc.org
MIX
Papier | Fördert
gute Waldnutzung
FSC® C083411

Kolibri 360 GmbH
Ferdinand-Jühlke-Straße 7
99095 Erfurt, Deutschland
produktsicherheit@kolibri360.de